Ecoscenography

"*Ecoscenography* addresses the urgent challenge of sustainable theatre practice from the perspective of scenography. Using a wealth of global examples, Beer examines the potential for scenography that is aligned with an ecological agenda. Through her own pioneering practice, she shows how ecological thinking challenges traditional modes of production. She proposes new approaches to making and viewing performance within a broader understanding of human and more-than-human relationships and shows how materials are co-creators in performance."

—Dr Joslin McKinney, *Associate Professor in Scenography, University of Leeds, UK*

"Beer's book, *Ecoscenography*, is a 21st Century hopeful response to previous generations of theatre making in a capitalist paradigm. She articulates the need for the theatre artist and artist/makers in general to bust out of technocratic thinking, move beyond the quantitative analysis of how much ruin we are creating or avoiding, and 'widen our identity' to include ways of knowing, seeing, being, and making that are relational to the complex ecosystem in which we live and work. This book is well-crafted response and necessary now most especially because she gives us hope that we are not lost but that art can actually transform us as individuals and us as a society. 'Hooray!' for hopeful thinking and 'Hooray!' for Tanja Beer!"

—April Viczko, *Associate Professor, University of Calgary & Project Lead, World Stage Design 2022*

"This statement from Tanja Beer's *Ecoscenography: An Introduction to Ecological Design for Performance* is very clearly summing up our inevitable situation: 'We are standing at the precipice of a burgeoning movement, one that has the potential to drastically change the way we make theatre in the years to come.' We are indeed in the midst of a serious and exciting change where many artists are realizing that a different approach is needed to start healing the planet. This book reveals many different possibilities to create sustainably, with communities, places, environments and 'engage in acts of care'. Tanja Beer's writing and her own practice inspire, energize and empower all readers to open up to ecological consciousness, make it core of their own collaborations and connect in building a holistic sustainable culture of creators."

—Markéta Fantová, *Artistic Director, Prague Quadrennial of Performance Design and Space*

"With *Ecoscenography: An Introduction to Ecological Design for Performance*, Tanja Beer provokes and encourages all theatre makers to embrace the transformational potentialities of a sustainable ethic. More than a guide, this book celebrates scenography as an agency for 'worlding-with', offering remarkable insights and tools for theatre making in the 21st century. As a performance designer, I clearly sense that there is no possible scenography other than Ecoscenography!"

—Dr Carolina E. Santo, *designer and Curator of Performance, Prague Quadrennial of Performance Design and Space 2022*

"Tanja Beer makes a compelling case for the need to embrace ecoscenography as an exciting new field that can contribute to the healing of our societies and our planet. With beautiful examples of ecological design in action, the book is a must that should be on the reading list of every theatre student and theatre maker."

—Chantal Bilodeau, *Artistic Director of The Arctic Cycle & Co-founder of Climate Change Theatre Action*

"In *Ecoscenography*, Tanja Beer takes the reader on a fascinating journey from ecological thinking towards sustainable practice, demonstrating how sustainability—far from being a limitation—opens up a new world of creative opportunities. Stage designers, but also theatre scholars and other practitioners in the field of performing arts, are introduced to an expanded concept of scenography, where the ideas of innovation, high-quality aesthetics, inspiration, ingenuity, become enriched and revitalised by an ecological ethic."

—Iphigenia Taxopoulou, *General Secretary of mitos21 European Theatre Network*

"If you are a designer, scenographer, artist or somebody that cares deeply about the future of our world then this is a book for you. Beautifully written, this thought-provoking introduction to Eco-Scenography introduces ecologic design solutions that are urgently needed. It is inspiring and exciting and deeply considered work that I highly recommend be a staple in education and on your personal bookshelf."

—Eliza Sweeney, *designer and eco-drama therapist*

"Tanja Beer's *Ecoscenography: An Introduction to Ecological Design for Performance* is a field-defining book for performance scholars and practitioners alike. Beer coined the term Ecoscenography in 2012 and the book offers a lucid, accessible and inspiring account of both the theoretical and practical dimension of ecoscenographic approaches. Accompanied by exquisite illustrations, the book charts the conceptual underpinnings of Ecoscenography—'ecological thinking', the history of sustainable theatre design, finishing with a detailed description of Beer's own framework for ecoscenographic practice. This is an important book for designers, performance-makers and anyone working in the performance sector interested in responding to climate emergency."

—Dr Kathryn Kelly, *Lecturer in Drama, School of Creative Practice, Queensland University of Technology*

Tanja Beer

Ecoscenography

An Introduction to Ecological Design for Performance

palgrave
macmillan

Tanja Beer
Queensland College of Art
Griffith University
Southport, QLD, Australia

ISBN 978-981-16-7177-7 ISBN 978-981-16-7178-4 (eBook)
https://doi.org/10.1007/978-981-16-7178-4

This Palgrave Macmillan imprint is published by the registered company Springer Nature Singapore Pte Ltd.
The registered company address is: 152 Beach Road, #21-01/04 Gateway East, Singapore 189721, Singapore

For Schatzi, with love
and
For all performance designers and theatremakers, past, present and emerging, with gratitude

Foreword

Being an environmentalist for as long as I can remember, considering the ecological impacts of my work has been a long-established value. Yet opportunities to practice ecologically have also been hard won in a discipline that is resistant to change. Chances to collaborate with this mindset have come slowly, while the urgency to act increases year-on-year. I have the advantage of my privilege as a cishet white man with ties to two wealthy nations. Relative wealth and mobility have allowed me to participate in the foundation of a sustainable theatrical design practice, including the creation of the Centre for Sustainable Practice in the Arts in 2008. But I, and any other individual scenographer with similar advantages, serve as mere novelties within the larger world of performance design if collective action eludes us. This is why this book is necessary: our survival on this planet requires all of us to engage with our environmental future, and this volume provides guidance and instruction to drive this change amongst all performance makers.

Have you noticed the changes happening around you? Can you sense the urgency of action? Is it the strength of hurricanes each season? Is it the massive wildfires across Australia, pacific and western Canada, or California? Or might it be something less apocalyptic.

In 2012, I moved to Toronto from Los Angeles to take on the unique position of Professor of Ecological Design for Performance at York University. My first winters here were grim: sub-zero for months, with snow and ice storms sweeping in with regularity. Sidewalks were single-file canyons of cleared snow for weeks. Those since have seemed milder. Snowfall hasn't fallen off precipitously, but we've not seen persistent accumulation. Streets are now rarely lined with snowdrifts from the plowed roads for very long. In these years—under a decade—the reality of climate change has become clear. Milder winters in Toronto have been a subtle person signal, but the explosive global events of this time period show us that climate change is here, and we must act—as citizens and professionals—to tackle it.

Movement in the performing arts is underway. I first met Tanja Beer at the Earth Matters on Stage Festival and Symposium hosted at Carnegie Mellon University in 2012, after coming to be aware of each other's work a little time before that. There were few theatrical designers working on sustainability issues at the time, and it was thrilling to know that there was someone also concerned about this topic, even if they were working on the opposite side of the planet. The following year, I served as the coordinator for the sustainability focused People, Planet, Profit track at

World Stage Design in Cardiff, where Tanja and I had the chance to work together formally. In 2016, I traveled to Australia for the Performance Studies International conference in Melbourne and a pre-conference meeting of the Performance+Design Working Group in Tasmania. Tanja was also involved in the working group, and we collectively programmed a sequence of panels on Expanded Scenography at the conference. Looking back, these events were at the beginning of the ecological turn in the performing arts, and the momentum is now building swiftly.

You may ask why sustainability and ecological consciousness matters for performance makers. Isn't our contribution to biodiversity loss and climate change negligible? The answer—at least for me—lies in the function of art, and especially performance, in society. It is how we construct and share our identities, both individually and collectively. Performance serves as a modelling exercise in which we gather to imagine society through a variety of lenses with the potential to envision other ways of seeing and being. If our current ways of living are not sustainable, we must engage in this practice to reconfigure our relations for survivability. Later in this book, Tanja states that 'ecological aesthetics asks us to consider how something can possibly be perceived as beautiful if it taxes and degrades living ecosystems'. My own position has coalesced with this ethos, and today I see performance through this prism. And so, what excites me about this book is that it provides a road map to rebuild our practices to serve an ecological purpose, and to reconfigure our relations to each other and to the non-human.

To close, I will commend this book to you by highlighting one more element. In Chap. 4, you will come to the section on co-creation (the start of the Ecoscenographic cycle), where you will find the idea of 'worlding-with'—the notion of co-creating mutually beneficial more-than-human partnerships. I sincerely hope this book will function as an act of co-creation and worlding-with for you.

York University, Toronto, ON, Canada Ian Garrett
January 2021

Preface

Like many stage designers, my interest in sustainability emanates from the scale of the environmental challenges our society faces and the desire to contribute to a positive future. Looking back, it is difficult to pinpoint a single moment that ignited my passion for ecological practice. Was it seeing my set thrown in a skip one too many times? Or perhaps feeling physically ill from the toxic fumes inhaled while expending several cans of spray paint on a single prop? Or being confronted by Al Gore's *The Inconvenient Truth* and its uncompromising depiction of the threat climate change represents to current and future generations? No doubt each of these events has played a role.

I have been passionate about environmental and social issues for as long as I can remember. I grew up with a strong desire (and indeed, privilege) to incorporate ecological considerations into my day-to-day life: buying food from local markets, taking public transport, avoiding single-use plastics, turning off lights, recycling and composting. However, when I walked into the theatre, all these considerations seemed to fall away. Somehow theatre gave me a licence to do the things I would never do at home. Perhaps it was because I was never taught to critique theatre practices from an ecological perspective, or query the consequences of my design intentions. Decadence was not questioned in my era of design education; on the contrary, it was encouraged if the budget would allow for it. We were trained to work towards Opening Night. How we got there, or what happened to our sets and costumes after the production ended, was neither a priority nor a consideration. I simply did not know how to integrate ecological concerns with my creative endeavours.

This changed in 2008, when I had the opportunity to work as an exhibition designer at one of Australia's major museums. The job came with a unique selling point: a team of ecologically conscious designers who were keen to put sustainability at the heart of what they did. Part of my job was to conduct research into sustainable strategies and products—a challenging task for a designer with no prior knowledge or experience in this area. It was not long until I was confronted with the uncomfortable realisation that my values were at odds with my design practices. Suddenly, I had deep concerns about the materials, products and life-cycle of my sets and costumes. I realised that despite only scratching the surface on the sustainability agenda, I already knew too much. There was no turning back.

In 2008 there was very little material available on the subject of ecological design for performance. I clearly remember searching the Internet and being dismayed at the lack of resources on the subject. While I found material on sustainable practices from other design fields, there seemed to be a gaping hole when it came to theatre. Why weren't we thinking about the consequences of what we design? Why had I myself only started to question it now? This questioning sparked a series of investigations into my own practice and finally the commencement of my PhD thesis on the topic three years later, in 2011.

This book provides an introduction to ecological design for performance, outlining my own research in the field over the last decade and showcasing the burgeoning body of work being produced by ecological theatre makers around the world. I am, of course, not alone in engaging with this topic. In recent years, sustainability has emerged as a significant part of many high-profile performance platforms—both in programming and in practice—across the globe. Pioneers in ecological theatre practice are demonstrating how the performing arts can be a unique and powerful platform to imagine and inspire new realities, with productions designed on sustainable principles demonstrating new approaches and artistic insights.

My motivation here is to synthesise the field for both established theatre makers wanting to engage with the topic and the next generation who—my own experience tells me—are yearning to contribute to their rapidly changing world. As we collectively face the reality that our practices have consequences, we are also presented with an opportunity to remake our profession. This 'ecological turning' is a pivotal point in the history of the performing arts, to be defined by the theatre makers of today, but particularly by those of tomorrow. The aim of this book is to catalyse ecological practices in performance design, providing a foundation from which new practices, new aesthetics and novel thinking can emerge.

The urgency for this reimagining has never been greater. Climate change, mass species loss, drought, forest felling and acidifying oceans pose immense threats to our environment, our society and our culture. In *This Is Not a Drill: An Extinction Rebellion Handbook* (2019), climate activist Dougald Hine summed up this era as, 'times in which impossible things happen'. Hine's prophetic words were at the forefront of my mind as I sat down to write this book in early January 2020, when my adopted homeland of Australia was literally on fire. By the end of what is now known as the 'Black Summer', wildfires had burnt 19 million hectares, destroyed 2779 homes and killed 34 people. An estimated one billion mammals, birds, reptiles and amphibians perished, and the fires contributed more CO_2 to the atmosphere than Australia's annual emissions. Smoke turned the sky red in New Zealand and blanketed its glaciers in smoke and ash. The fires—driven by unprecedented drought and heat—bought home the urgency of the climate crisis and the need to reshape our relations with the environment that supports us. I strongly believe that art and culture have a powerful role to play in this reformation.

We also stand at a unique moment in time to pursue an ecological turn in the performing arts. The last 12 months, spanning 2020/2021, has bought with it a global pandemic that has enforced a pause in our practices. Just as there have been calls to re-start our economy with a green revolution, so too can we re-start theatre practices with a new ecological ethic. The pandemic has also brought home the inherent inequality that underlines and intersects with our climate crisis, demonstrating that the ecological is social and political. We must convene to tackle this moment in our history with humility, gusto and resourcefulness.

For me, the underlying spirit of this movement is one of opportunity. Enforced sustainable practices infer the negative; they suggest unwelcome constraints. On the contrary, I am interested in how the coming decades can be framed as the time of potential, an era when theatre artists rise to the challenge of what it means to bring an ecological perspective into their work, not because they are forced to, but because they embrace the transformative potential that it brings. Related industries, such as architecture, product design and fashion, have already shown us how an ecological ethic can reap enormous rewards.

It is my hope that this book will help reveal these opportunities to performance designers and theatre makers alike and, as outlined above, provide a foundation from which ecological practices can build in the coming decades. There is already more than a whiff of revolution in the air. Sustainability with a capital 'S' is finally here. The time is now to seize the potential that this new era in the performing arts brings.

Southport, QLD, Australia Tanja Beer

Acknowledgements

First and foremost, I would like to express my deep gratitude to the Traditional Owners of the many Countries that make up Australia, upon whose lands I have had the benefit of writing and conducting my research and creative projects. Particularly, the people of the Kulin Nation, of the Woi Wurrung and Boon Wurrung language groups and the Yugambeh language group of the Gold Coast region, where much of this book was written. I recognise that all constructed environments and cultural activities in Australia take place on First Nations Country, which have histories of more than 60,000 years of careful custodianship. I acknowledge that Australia was founded on the genocide and dispossession of First Nations people, and sovereignty was never ceded. Australia always was, and always will be Aboriginal land. I believe that a core part of engaging with sustainability begins with recognising the violent colonial mindsets and actions that have depleted our intimate connections with living systems. Indigenous Australian approaches to environmental custodianship provide vital inspiration and guidance when considering how we co-create cultural practices that are in tune with living systems. First Peoples have long demonstrated how the ecological is cultural and that the merging of art, land and story must be part of our collective healing. We must make space to listen.

This book would not have been possible without my PhD and postdoctoral research undertaken at the University of Melbourne from 2011 to 2019. I am grateful for all the support that I received during my time at the Faculty of Architecture, Building and Planning, which shaped my thinking and artistic practice deeply. My extraordinary supervisor and mentor Dominique Hes was vital in developing my understanding of sustainability and introducing me to the foundations of ecological design thinking, which underpins this book. These concepts have been further enriched by the wonderful Janet McGaw, who welcomed me into the world of architecture with open arms and helped me to expand my philosophical readings. I am also grateful for my PhD supervisor Alyson Campbell (Victorian College of the Arts), whose acute attention to detail and much-needed objectivity provided me with valuable feedback throughout my thesis journey. Thank you also to the marvellous Sally Mackey, who was my 'exchange-supervisor' during my time at the Royal Central School of Speech and Drama (London, 2013/2014) and my erudite and benevolent PhD reviewers Dorita Hannah and Raymond Cole, who provided much-needed critique along my research journey.

Learning has been a big part of my life, and I am equally grateful for my time at the Universität für Musik und darstellende Kunst Graz (1995–2000) and the Victorian College of the Arts (2001–2002), where I first learnt the art of performance design and theatre making respectively. More recently, I could not have completed this book without the support of my new colleagues at the Queensland College of Art (Griffith University). I am also grateful to all my fellow academics and thought-provoking students who helped expand my viewpoints and challenged and nurtured me from the sidelines over the last decade.

Writing a book is a solitary process, yet we do not write alone. I have been blessed with a supportive community of writers and thinkers who have helped bring this book forth. Thank you Aby Cohen, Iphigenia Taxopoulou, Hannah Harding and Ashlee Hughes for all your support and feedback during this process, and especially, to Ian Garrett and Rachel Hann for their thoughtful and valued contributions to this book. I am particularly grateful for my collaborator Lisa Woynarski, whose writings in the fields of performance and ecology continue to inspire my thinking and practice. A special mention must also go to my research assistant Brittany Laidlaw, who undertook the painstaking task of compiling and documenting all the references for this text.

I am tremendously grateful to have had the opportunity to work with my exceptionally talented sister, Gisela, who created all the illustrations for this book. Thank you Gisela for all your incredible work in making Ecoscenography come alive through your beautiful drawings—I am so in awe of your artistic abilities. Thank you also for helping me with the Index!

I have been fortunate to share many of the concepts of this book across multiple platforms, both internationally and nationally. Thank you to all the members of the Ecoscenography Facebook Group, who have helped provide a platform for ecological thinking in performance practice. I would also like to express my sincere gratitude to the Melbourne Museum, Julies Bicycle, the Society of British Theatre Designers, the Prague Quadrennial, World Stage Design, *Theatre and Performance Design Journal* and the IFTR Scenography and Architecture and Theatre Working Groups, who have provided a rich platform for discussing ideas and connecting with fellow eco-minded designers and theatre makers over the years. A special mention must go to all the wonderful thinkers who have supported and enriched my work over the last decade: Pamela Howard, Wallace Heim, Una Chaudhuri, Sofia Pantouvaki, Jane Collins, Joslin McKinney, Scott Palmer, Thea Brejzek, Chantal Bilodeau, Jeremy Pickard, Lanxing Fu, Annett Baumast, Angela Campbell and Michael Mehler. Thank you for your generosity, wisdom and writings.

I am exceedingly grateful to all the extraordinary artists and thinkers that are featured in the book, especially those that have shared their insights and photographs of their work with me, especially: Noémie Avidar, Andrea Carr, Annike Flo, Ingvill Fossheim, Ian Garrett, Marie-Renée Bourget Harvey, Mona Kastell, Silje Kise, Thierry Leonardi, Aris Pretelin-Esteves, Janne Robberstad, Imogen Ross, Ruth Stringer, Xiao Ting and Donyale Werle. Their resilience and creativity continues to inspire and give me hope that we are indeed on a path to a more ecological practice.

I am indebted to the hundreds of wonderful collaborators who have worked with me on many ecological projects across the world, especially on the various iterations of *This Is Not Rubbish*, *The Living Stage* and *Running Wild*, the credits for which are featured at the back of this book. In particular, I want to thank my longstanding eco-collaborators Jennifer Tran, Ashlee Hughes, Pia Guilliatt and Cristina Hernandez Santin for their problem-solving prowess and patience, and for always having my back.

I owe a debt of appreciation to my parents Brigitte and Gernot for their continuous encouragement in my quest for learning, as well as instilling a love for art and ecology throughout my childhood. Thank you for your love and support throughout my career as well as giving me the confidence and freedom to follow my creative dreams as a young girl. Thank you also to Armin and Christa, who not only supported me during my educational foundations in Austria, but have always nurtured my artistic and academic pursuits with such love and encouragement. My Austrian heritage, with its strong affinity with nature and culture, is still one of the greatest influences on the work I do today as an ecological designer and community artist.

My eternal gratitude is also extended to Milton, who has been an incredible support throughout my career, from filling sandbags and lugging apple boxes during the installation of the first Living Stage, to being the dedicated 'plant waterer' on almost all subsequent Living Stage projects, providing financial support when the kitty was running low, and finally, to proofreading this entire book. I cannot thank you enough Milton for your love and dedication. I am so grateful to have you in my life.

Finally, to Geoff who has nurtured this journey from the sidelines on a daily basis over the last eight years, who graciously listened to my ideas and provided much-needed editing support, and most importantly, whose work as an ecologist and deep connection to the natural world continue to inspire me to be a better person and citizen of the earth.

A Brief Outline of the Book

This book provides an overview of the philosophies and practices that encompass my conceptualisation of ecological design for performance. It begins with an *Introduction*, which summarises the status of sustainability in the performing arts, particularly how it has evolved in recent years—from early developments of environmental consciousness in the 1990s, to more recent shifts in ecologically engaged theatre making and design. Tensions and opportunities of sustainable production are interrogated, revealing new insights for responses to the environmental challenges of the coming decades. Drawing upon contemporary understandings of the field, I introduce the concept of 'Ecoscenography', the fundamental tenets of which are investigated in the subsequent chapters of the book.

Chapter Two: Ecological Thinking outlines the conceptual and philosophical underpinnings of Ecoscenography and explores ways in which theatre makers can consider their practice within the broader system of ecological organisation. By reviewing literature from a variety of fields across systems thinking, environmental humanities, aesthetics and eco-materialism, I examine how expanded ideas of material entanglement can be integrated with the ephemerality of scenographic practice to create work that encompasses environmental, social and political potential.

Chapter Three: Ecological Practice draws upon a range of ecological design theories and precedents and explores their application to

Ecoscenography. The chapter examines ways in which theatre makers might move beyond eco-efficiencies or 'less bad' consequences to integrate design processes that are in tune with living systems. It includes a broad range of tangible examples across a range of fields to highlight the potential of sustainable practice. While there are challenges and barriers to implementing sustainable practice, I argue that Ecoscenography opens up avenues for exploration in both conventional practice and more expansive areas of scenographic activation, encompassing novel ways through which scenographies are designed, constructed and distributed.

Chapter Four: Ecoscenography in Three Acts proposes three stages of Ecoscenography—co-creation (pre-production), celebration (production) and circulation (post-production)—which are considered equally fundamental to the aesthetic consideration of the work. The chapter provides examples of scenographers who have pioneered new practices and aesthetics for a sustainable future and introduces the Ecoscenography trajectory, outlining challenges and opportunities of practice.

Chapter Five: Two Case Studies in Ecoscenography introduces examples from my own practice—*The Living Stage* and *This Is Not Rubbish*—where the possibilities of Ecoscenography are explored beyond the confines of traditional theatre practice. The chapter is to be read in conjunction with a selection of digital recordings of the creative work, which can be accessed via my website www.ecoscenography.com. I close by considering ways forward for Ecoscenographic futures.

The *Conclusion* offers a concise summary of the key arguments of the book drawing out implications for scenographic scholarship and practice. I contend that considering wider socio-ecological factors of performance design leads to a renewed investigation of scenographic materials, processes, aesthetics, roles and partnerships that can guide a positive future.

As well as featuring images of ecological design projects from multiple artists, the book also includes a series of illustrations by Gisela Beer, which seek to depict key concepts of Ecoscenography and offer a visual perspective to the text. Their intention is to bring the concepts and theories of Ecoscenography to life. The use of the 'bee and the sunflower' to explain the cycles of Ecoscenography is our combined concept.

In addition to the book, a selection of interviews, photographic and film documentations are available on my Ecoscenography website (www.

ecoscenography.com) to complement the writing and provide additional ways of articulating and evidencing the work. While I am aware that the reader cannot experience the referenced 'scenographic moments' as they were originally conceived, executed and activated in their original physical and spatial realm, my hope is that the corresponding digital media will allow readers to comprehend the projects in greater depth.

Clarification of Terms and Scope

Many of the terms that I use in this book can be interpreted in numerous ways, some of which are contradictory. As such, it is important to clarify the meanings here. Firstly, my conceptualisation of scenography refers to the interrelationships between 'performers, space, object, material, light and sound' (following Palmer 2011, 52). Thus, scenography encompasses sound, light and costume under the banner of spatial design. Nevertheless, my focus in this book has largely been on the 'set and costume design' elements of scenography, in keeping with my practice over the last 20 years. I acknowledge that this is somewhat of a restrictive lens, but the intention has been to focus on my area of expertise and to bring in other examples where I could.

I refer to the term 'designer' and 'scenographer' interchangeably. While these terms can be considered separately (e.g., in Howard 2009), they are more often paired together in practice (Collins and Nisbet 2010, 1). In Australia, the term 'scenographer' is rarely used outside of academia and is commonly replaced by the term 'designer'. As this book references broader perspectives of design from the field of built environments and sustainability, the use of the term 'designer' is also helpful as it works across disciplinary fields and thus breaks down perceived binaries. It also encompasses the term 'performance design', which as a 'porous, fluctuating term—works across a broad spectrum that embraces the theatrical, historical and quotidian' (Hannah and Harsløf 2008, 12). While this book is written from the perspective of a performance designer, I use the term 'theatre artist', 'theatre maker' or 'performance maker' to include any practitioner who works within the performing arts (who directly or indirectly engages with scenography).

I also use the terms 'performance', 'performing arts' and 'theatre' interchangeably. By this, I am incorporating all aspects of performance, including (but not limited to) text-based theatre, dance, opera, music theatre, physical theatre, circus and the more expansive and unconventional realms of live art and/or performance art.

The term 'sustainability' and 'sustainable design' have been frequently redefined and contested. Here, I use 'sustainability' to mean the continuing relationship between humans and ecological systems, where 'sustainability integrates natural systems with human patterns and celebrates continuity, uniqueness and placemaking' (Early 1993). I also incorporate Margaret Robertson's approach to 'sustainability' both as an idea and as discipline:

> 'Sustainability' as an idea refers to the body of knowledge that deals with how dynamic systems work on this planet we call home, including what we know about the current health or decline of those systems. 'Sustainability' as a discipline refers to humanity's rapidly-evolving response to the urgent planetary challenges we all face, a response that includes emerging professional opportunities. (Robertson 2014, 3)

I use the terms 'ecology' and 'ecological' to mean the integration of environmental, social, political and cultural aspects. In some cases, the term 'socio-ecological' is used to draw attention to the social aspects of ecology. 'Ecological design' and 'sustainable design' are often used interchangeably; however, I tend to use 'ecological design' when referring to practices that explicitly integrate ecological or relational thinking in their work (further discussed in Chaps. 2 and 3).

A key priority in writing this book has been to ensure its accessibility to a range of readers, including theatre practitioners and researchers. It is my hope that this book might act as a source of contemplation and inspiration as to how an ecological approach to scenography and production can be pursued—both in academia and in practice—with a focus on integrating critical and holistic ideas of sustainability with real-world examples. As the concept of Ecoscenography is still in its infancy, I have opted for an introductory text that aims to sketch out the potential of the field in its early stages of theoretical and practical development. It is by no means comprehensive but a starting point for further dialogue and discussion.

While I have endeavoured to cover a range of sustainability perspectives in the performing arts, my access to writings and precedents has been largely limited to the Global North, which I readily acknowledge does not account for the full array of artists and commentators working in this area. I am a white educated cisgendered woman of European heritage, who has been fortunate to grow up in a privileged middle-class society. As such, I am aware that this book is written from one particular perspective and does not cover the full spectrum of voices and experiences that are part of this field. I am deeply passionate about social-ecological justice and also accept full responsibility for any unintentional errors that I will undoubtedly make in the writing of this book. While I have attempted to diversify my sources and precedents, it is important to note that this book provides only a glimpse into the possibilities of this rapidly growing field. It is my hope that as the intersection of theatre and sustainability gains currency, we will see a more substantial and diverse picture of what this entails across cultures.

References

Collins, Jane, and Andrew Nisbet. 2010. *Theatre and Performance Design: A Reader in Scenography*. London, UK: Routledge.

Early, D. 1993. *What Is Sustainable Design*. Berkley, CA: Berkley: Society of Urban Ecology.

Extinction Rebellion. 2019. *This Is Not a Drill: An Extinction Rebellion Handbook*. London, UK: Penguin Random House.

Hannah, Dorita, and Olav Harsløf. 2008. *Performance Design*. Copenhagen, Denmark: Museum Tusculanum Press.

Howard, Pamela. 2009. *What is Scenography?* Oxfordshire, UK: Routledge.
Palmer, Scott. 2011. Chapter 3: Space. In *Performance Perspectives: A Critical Introduction*, eds. Jonathan Pitches, and Sita Popat, 52–81. Basingstoke, UK: Palgrave Macmillan.
Robertson, Margaret. 2014. *Sustainability: Principles and Practice*. London, UK: Routledge/Taylor & Francis Group.

Contents

List of Figures

1

Introduction

T. Beer, *Ecoscenography*, https://doi.org/10.1007/978-981-16-7178-4_1

As the world enters a new age of environmental uncertainty, the performing arts faces a choice: it can embrace the ecological turn sweeping across society or it can shirk the challenge. The choice is not merely a moral or ethical one, but one that has significant implications for the industry's relevance, vitality and innovation through the twenty-first century. In her speech at World Stage Design in 2013, eco-arts scholar Wallace Heim argued that the time will soon come when theatres will need to justify excessive and unsustainable behaviour—when 'those who want massive spectacles, world tours, and blazing lights will have to openly justify and account for those technologies and excessive and exceptional drains' (2013). Of course, this shift is already underway across our global society; conventional practices modelled on the unethical and unsustainable realities of the modern era are fast becoming outdated. Priorities are shifting to consider practices that are not only environmentally conscious, but also more culturally diverse and inclusive. Things that seemed 'acceptable' a few years ago—disposable coffee cups, single-use plastic, meat-heavy diets and frequent air travel—are becoming increasingly unacceptable. Hastened by the pace of climate change, our prevalent mechanistic, imperialist and rationalist worldview, built upon a mandate of progress, capitalism and modernity, is rapidly coming undone. As Naomi Klein (2014, 21) writes, we have been 'at war' with the very life that sustains us.

In this age of environmental awareness, I argue that we need a new philosophy for theatre production that promotes ecological, holistic, interconnected and symbiotic practices. This new philosophy will be a seismic shift, at least in some quarters. As highlighted above, scenographers—including myself—have been trained to work towards Opening Night. How we 'get there' or what happens to our sets and costumes after the production ends has been neither a priority nor a consideration. Our focus as designers has typically been to create 'experiences of impermanence'—often extravagant spectacles with little regard for the prevailing permanence of unwanted remains. This emphasis on the new and disposable in theatre practice has largely mirrored the consumptive habits of the modern era. Up until the 1970s, the reuse of scenery and stock items in large theatre companies was still standard practice with scenic drops, stairs, flats and furniture being repainted or repurposed to provide new opportunities for design (Morris 2007, 29–30). However, by the mid-1980s, theatre companies were throwing away scenery 'as if it were an investment that had lost all its glimmer', contributing to the tons of demolition waste that is most prevalent in our landfills (1–2, 43).

It is true that the environmental impacts of the performing arts do not match those from other industries; however, theatre's prominent 'take, make, dispose' attitude is fast becoming outdated. Related industries such as architecture and fashion have been steadily accruing a long list of environmental resources, tools and precedents over the past decades. In contrast, sustainability has continued to be a niche (albeit growing) concern amongst theatre makers, both those in practice and those in academia. The environmental impacts of theatrical design are yet to be adequately and widely documented, nor has there been well-researched attempts to examine means by which current theatre practices can be reimagined to comply with ecological principles (Morris 2007; Brunner and Mehler 2013; Omen 2014; Peeters 2012). Ecological design for performance is arguably the most challenging and least investigated component of sustainability (Brunner and Mehler 2013, 32), and the perception remains among many that 'sustainability and theatre do not mix'.

This book seeks explicitly to break this assumption and –responding directly to Heim's call in 2013—explore what it might mean to embrace an ecological ethic in performance design, not as a limitation but as an opportunity. In addition to the environmental imperatives, I also seek to define this opportunity in terms of the continued rise of scenographers as agents of change. Recent machinations in the field have rendered an increased captivation with what 'scenography' *does*, rather than what it *is* (McKinney and Palmer 2017, 19). Scenography as a verb has opened up our discipline to one which embraces not only the things that make up performance, but also the interrelationships between things, beings, places and communities (Beer et al. 2018). While the early decades of the twenty-first century have been preoccupied with defining and redefining of scenography, I propose that over the next two decades, we must pivot to further interrogate the agentic capacities of our field, not only in terms of the wondrous 'worlds' and 'experiences' that we create for our audiences, but also in terms of the ecological, social and political consequences, impacts and messaging behind our work. I view the pursuit of an ecological ethic as an opportunity for scenographers to continue their evolution to being artists and thinkers in their own right. Artists that critique the past and propose the future, artists that have impact well beyond the confines of the theatre.

This chapter provides an introduction to the burgeoning field of theatre and sustainability. It 'sets the scene' for the emerging concept of Ecoscenography—a neologism that I use to encapsulate the integration of an ecological ethic with performance design. The intersection of design and sustainability traverses many disciplines and foci in the performing arts. This chapter provides an overview of the context, background, dialogue and perspectives that underpin this founding concept across theatre production, performance making and scenography. While sustainability in theatre and performance design is growing, there is still a need for more diverse scholarship and practice engaged in ecological thinking across a range of ethnic, racial and geographic contexts.

Setting the Eco-scene

The emergence of a sustainability movement for theatre production can be traced to Larry K. Fried and Teresa J. May's *Greening Up Our Houses: A Guidebook to an Ecologically Sensitive Theatre Organization*, published in 1992. Fried and May sought to educate the performing arts community on its increasingly wasteful practices and inspire an ecological ethic that would jolt the industry out of environmental complacency. Nevertheless, while the work generated considerable academic interest, it largely failed to shift the industry, which predominantly remained oblivious and unwelcoming of ecological concerns during the 1990s (Butler 2009, 110). In a pointed discussion of the issue more than 15 years later, theatre maker and scholar Damond Morris (2007, 2, 43) concluded:

> The theatrical industry is broken, plundering the earth of valuable resources without a thought for the wellbeing of future generations. While other industries are taking on board green practices, the theatre industry is painfully unaware...In the 20th century there were, modestly, hundreds of thousands of productions that moved through millions...of tons of waste.

> This waste, classified as construction and demolition waste, is one of the most prevalent materials in our landfills. To curb the waste we create in the theatre industry we must find new ways of thinking and new ways of working without it.

Stinging critiques, such as those of Morris, have raised the profile of sustainable practice in the performing arts over the last decade, but it is worth interrogating why the performing arts did not progress sooner on the environmental agenda. At a green theatre event that I attended in Melbourne in 2012, negative preconceptions of sustainability—'expensive', 'boring', 'time-consuming' and 'incompatible with high-quality aesthetics'—were identified as the main contributors to environmental complacency. Another contributing factor may have been the fact that the theatre industry has been far too preoccupied with its own battles to acknowledge, or tackle, its effect on the environment (Butler 2009, 110). Many theatres have struggled for years to merely survive economically, even contemplating environmental sustainability bordered on hubris (Stancato 2010, 36). However, there was also a sense of entitlement and complacency in some sections of the performing arts with regard to environmental concerns. As performance artist Mojisola Adebayo (2009) described in her interview on *The Ashton Directory* website:

> Some people in theatre believe they are above climate change…that theatre is inherently good for you and, therefore, nothing in our work could be harmful. The attitude that we are above this subject needs to change.

Adebayo's comments were founded in the larger societal shift in ecological awareness that was developing in the first decade of the twenty-first century. Broader media coverage (such as those that were highlighted at international climate change summits) and documentary films such as Al Gore's *An Inconvenient Truth* (2006) presented a call for action that has begun to filter through to even the most reluctant sectors and industries. Bill McKibben's essay, 'What the warming world needs now is art, sweet art' (2005), was widely shared and cited, giving birth to a series of ecological artworks. In her second edition of *What Is Scenography?* (2009), renowned scenographer Pamela Howard drew attention to theatre's unsustainable habits, claiming that performance makers:

> have a responsibility in these times to address the problems of today and to fuel change and alternatives…to show that rich and beautiful theatre can be made without creating mountains of waste…scenic construction that can rarely be used again does not set a good example to spectators who diligently recycle bottles and newspapers, and grow their own vegetables. (Howard 2009, 216, 222)

In 2008, two major campaigns were independently announced for reducing environmental impacts in the performing arts: the Mayor of London's Green Theatre Plan and the Broadway Green Alliance campaign. Many organisations[1] worldwide have followed suit, offering tools, case studies and seminars dedicated to greening the industry, with an increasing number of theatre companies and festivals developing sustainable production policies and strategies.

In addition to these initiatives, two major events have been particularly influential in spurring on the rise of a sustainable consciousness in performance design. In 2013, Ian Garrett—co-founder and director of the Centre for Sustainable Practice in the Arts (CSPA) and Associate Professor of Ecological Design for Performance[2]—led the first 'People, Profit, Planet' sustainability programme at the World Stage Design (WSD) congress in Cardiff, UK. The year before, Arts Council England became the first national arts funding body to 'legislate' to reduce carbon emissions, through its Environmental Action Plan. Since then, the sustainability movement in the performing arts has progressed rapidly, with companies such as the Sydney Theatre Company turning to renewable energy sources.

Response to ecological urgency surged in the 2010s, with a proliferation of international theatre festivals, conferences, symposiums, initiatives, books and articles focused on the intersection between sustainability, performance and ecology. Major events[3] included Performance Studies International (Performing Climates, Melbourne, 2016) and, more recently, the International Federation of Theatre Research conference (*Ecologies: Environments, Sustainability, and Politics*, Galway 2021). In

Australia, national initiatives such as Tipping Point, Greener Live Performances and Greening the Arts provided sustainability guides and carbon calculator tools for theatre production. The 'surge' of the last decade has kept pace with, and been fuelled by, the broader societal movement that seeks to change the course of human impact on our environment and climate. The dire warnings of the Intergovernmental Panel on Climate Change (IPCC), activism from Extinction Rebellion and prolific campaigning by youth climate activists such as Greta Thunberg, have brought environmental concerns to the forefront of mainstream consciousness. As Paul Brunner and Olivia Ranseen (2017, 20) contend:

> Today, we know more than ever about the human role in climate change, and a growing collection of literature examines the intersection of sustainability and theatre arts, further legitimizing this small pioneering field of study.

While sustainable theatre practices continue to encounter a number of challenges and obstacles (as I detail in the next section), the movement is now well established and growing. Scenic designer and sustainability advocate Sandra Goldmark (2017, 28) explains how:

> As designers and theatre makers, we must make it part of our job to think of our work in a larger context—we must think about the materials and objects we use, where they come from, what they are made of, where they will go after we use them, and the impact they have at each of these steps.

It is clear that sustainability as a key concern for theatre has arrived. The industry's 'ecological turn' has begun, and the twenty-first century will be one in which environmental concerns are at the forefront of our stories, our aesthetics and our designs.

The Design Problem

Theatre makers all over the world are heeding the call of environmentally conscious practices, yet notions of sustainability are still dominated by trepidations of creative limitations. Rather than embracing the possibilities of ecological challenges as part of their creative practice, even the most devoted sustainability advocates 'are focused on doing less harm, on using fewer resources' rather than exploring new forms of theatre making (Brunner and Mehler 2013, 32). As theatre artist Zoe Svensen (2019) writes:

> For so many of us living in cultures built on fossil fuel use, it is much easier to see losses and sacrifices than it is to see what might be gained by a transition to a world that faced up to the crisis.

The idea that artistic integrity cannot be compromised for a more sustainable production also reflects the opinion of many performing arts organisations who consider theatrical production to be inherently unconducive to sustainable practice (Morris 2007, 5). This has been escalated by the number of artists who have felt the need to be unencumbered by sustainability measures in order to have the freedom and independence to execute their creative vision regardless of its ecological consequences (Barnard and Briscoe 2015). As dramaturge Jeroen Peeters (2012) has observed:

> With slogans such as 'artistic autonomy' as talisman, makers appear to be ignoring the social-ecological crisis, and managers treat it simply as a case of efficient building management. Ecology is not considered a theme and field of research worthy of attention on its own, but rather is reduced to a necessary evil or altruistic engagement.

Contributing to this mindset is the fact that ecological thinking contradicts conventional contexts of theatre making that reflects wider socio-economic and mechanistic structures that privilege economic wealth over ecological vitality (Heim 2013). The 'urgency' of the social-ecological crisis has led theatre makers to implement efficient solutions 'without thoroughly analysing or calling into question the current system' (Peeters 2012, 86). Brunner and Mehler (2013, 26) make the implication explicit, arguing that embracing an ecological mindset requires that practitioners must first question theatre's unsustainable mode of production and consumption. Playwright Chantal Bilodeau (2019) emphasises how:

> Discrimination, economic and environmental injustice, and resource depletion are all manifestations of the same system gone awry. And to change that system, we can't just tinker with individual elements—we have to rethink the whole synergetic mess.

Resistance to a sustainable approach is also entrenched in the industry's appetite for spectacle and the need to continuously begin anew. No artistic director will tell their designers to limit themselves in order to reduce consumption and waste generated by their productions (Lawler 2008, 59). Current notions of sustainability, often limited to the principles of reduce, reuse and recycle, do not fit the model of extravagance to which the industry aspires. Many producers and directors encourage designers to create a visual experience that serves the audience's high-quality expectations: a world that is both sophisticated and entirely different from other designs seen previously (Morris 2007). Theatre's highly collaborative nature also requires a team effort in reaching sustainability goals. Sustainability entails a collective focus:

> One designer trying to be sustainable won't work if the director is not on the same page, or the rest of the team doesn't understand why. Directors, producers, and stage managers can become great leaders in this area as they are in charge of large portions of the production itself. (Cattenazzi 2019, 7)

While the above speaks to resistance to the embrace of an ecological ethic in the performing arts, it is also true that part of the sector's sluggishness to embrace sustainability results from inherent limitations. Even when there is openness to collaboratively explore creative and evocative approaches to sustainable design, theatre companies' tight budgets, precarious funding and short production schedules allow little opportunity for experimentation (Allen et al. 2014, 14). Larger sectors, such as architecture and fashion, have much more substantial budgets, produce longer-lasting designs and have more opportunities to repeat sustainable strategies across projects. In the theatre, every stage design is unique in every aspect, from its size and materials to the people involved, making it hard to standardise an approach. A stage designer's greatest priority is on creating a 'world' that supports the dramatic work. The use of sustainable materials and technologies is generally not perceived as part of the artistic purpose; it is not part of 'serving the play' (Garrett 2012, 201). Thus, designing for the stage includes a unique set of constraints that can make an ecological approach more problematic than traditional construction (202). As French sustainability consultant Thierry Leonardi explains with regard to the Circular Economy project OSCaR (Opera Sceneries Circularity and Resource efficiency):

> It could be said of the design and production of sets that they are an "economy of prototypes": no standardization and no mass production is possible. They are the result of artistic creation; they are the translation of a concept that could be said to be unique, linked to dramaturgy and a staging project. And they are protected by copyright. In addition…opera theatres have the objective of programming a single work on several occasions, over an indefinite period of time, which is often counted in years. This results in the need to store the sets; this storage induces a financial cost for theatres through rentals, energy and maintenance, and an environmental and financial cost for the regions (land use). (Leonardi 2018)

Time is also a major constraint for testing out new ways of doing, something with which the theatre industry struggles already (Brunner and Mehler 2013, 29). Designers are often forced into investing their own time, commitments and resources in experimenting with new materials and processes—often at the risk of receiving little acknowledgement or benefit for this investment from theatre organisations. Indeed, many are concerned that they may lose out on work prospects due to their commitment to sustainability being at odds with directors and producers. With designers frequently overworked and underpaid, it comes as no surprise that many professionals are still shying away from embracing (or even highlighting) ecological practice as part of their process. Incorporating sustainability in production meetings and discussions can 'require stretching out standard timelines' (Brunner and Mehler 2013, 30). In addition, designers are not usually responsible for what happens to their work after closing night; sets and costumes are often entrusted to the production manager, whose job is to decide if an item is destined for storage, reuse, recycling or landfill.

While there has been an effort to address the above issues, literature and case studies that investigate the deeper issues and potential of the subject are still scarce. In many cases, sustainable practice has been dominated by the idea that artists 'should get on with being creative' leaving the job of sustainability in the hands of production managers and technical staff (Barnard and Briscoe 2015). Without the adequate structures, tools and enthusiasm, many practitioners will find little opportunity to pursue sustainability.

Nevertheless, 'theatre cannot wait for sluggish institutions to take the lead, just like climate change cannot wait for governments to regulate or big corporations to smarten up' (Bilodeau 2015a). Challenging conventions is necessary for artistic practice to engage with more ecological views (Heim 2013). As Brunner and Mehler (2013, 26) argue, 'we cannot dismantle the system within which we work, but we can attempt to alter that system a bit at a time'. Subsequently, Brunner and Ranseen (2017, 22) contend that this does not need to be seen as a limitation:

> Theatre designers, production professionals, and technologists develop team oriented and innovative solutions, working under tight time constraints to clearly identify problems, seek solutions, and carry out ideas. As such, they are well prepared to envision new approaches to making art—solutions that can have profound and positive impacts on our communities. If theatre is to assume a leadership role as a change agent for sustainability, then it's important to understand how to better identify the reasons for resistance and discuss methods to overcome them.

Tony award-winning sustainability pioneers such as British lighting designer Paule Constable (*War Horse*) and American set designer Donyale Werle (*Peter and the Starcatcher*) have long demonstrated the potential of taking matters into their own hands, despite working within conventional circumstances of commercial and high-art sectors. For Constable and Werle (whom I feature later in this book), neither working with an ecological ethic is seen as a limitation, nor are sustainability and high-quality aesthetics seen as mutually exclusive. Working sustainably celebrates innovation and challenges them to think about what is at hand, as well as what is possible.

In summary, holding onto old paradigms has led to sustainable design being construed largely as one of 'limiting' or 'restricting' existing artistic practices. In reality this need not be the case, but the perception is one that must be overcome to progress ecological design for performance. Defining our lives in the negative—focused on what *not* to do or forcing old paradigms to be 'less bad'—will not motivate significant change (McDonough and Braungart 2002). Ecological action needs to be innovative, stimulating and deeply creative for the performing arts to embrace new ways of thinking that will overturn existing and intrinsically unsustainable modes of practice. Richard Heinberg, Senior Fellow of the Post Carbon Institute, sums it up simply: 'If it's not fun you're not doing it right' (Heinberg and Lerch 2010, 446).

Ecological Approaches to Performance Making

Sustainability initiatives in theatre production have played a vital role in advocating and supporting ecological change, but their focus has been largely pursued as an exercise in economic and technological efficiency—a 'technocratic' focus that has not integrated cultural potential. Combining arts and sustainability has been largely limited to 'an instrumental tick-box exercise' (Allen et al. 2014, 5). However, in recent years, the field has evolved from one in which sustainability has been behind the scenes, to one in which the stage has become a platform to engage theatre audiences with ecological issues. Theatre artists have increasingly called for the need to:

> promote sustainability in its full sense, enabling individuals and organizations to see that it is not simply about reducing carbon emissions but about more efficient and effective allocation of resources, meaningful interactions with communities, ideas and aspiration, social justice—it's about the society we want to be, it's about leading the discussion. (Evans in Allen et al. 2014, 29)

Artists exploring ecological issues 'can serve as a beacon' by helping us to 'weave together the fabric of values that will inform our future decisions' (Bilodeau 2015b). This cultural change is already underway:

> It's happening on theatre, concert hall, and festival stages; in museums, parks, and public libraries; at music labels and recording studios; and in the products and ideas of countless creatives, designers, artisans, experimentalists, and visionaries. (Latham 2019)

The ecological imperative has seen a number of artists turning towards unconventional forms of performance and devising hybrids (Heim 2013). For example, the Arts Council Wales Emergence Culture Shift report on sustainable arts practice (Allen et al. 2014) demonstrated how ecologically engaged artists are increasingly identifying a sense of civic responsibility in the way in which they practise.[4] The objective of these artists has progressively become one of creating 'relational', 'empathic', 'participatory', 'collective' and 'community-orientated' practices rather than exemplifying personal expression (Allen et al. 2014, 12–13, 27–28, 36). The focus of sustainability has shifted from individual artists or collectives, to embracing communities and giving them a central role in creative practice (Bilodeau 2015c). With landfills, ocean plastic and food waste becoming a growing concern of the twenty-first century, artists of all disciplines have increasingly embraced the potential of the discarded to counteract the unfettered capitalist 'myth of endless seamless progress' and 'continual creation of the new' (Edensor 2005, 101).

Not only are many independent theatre practitioners rethinking their roles as artists, they are also questioning what it means to be 'successful'[5] in response to the current global context (Allen et al. 2014, 14). Rather than following a 'traditional model' of success measured by flagship and high-profile venues, artists are increasingly 're-addressing or re-making' their own practices, letting go of former 'reputations or identities in favour of uncertainty and potential isolation' (Ibid.). Ecological-orientated artists 'often make new types of work, in surprising and unconventional spaces. They are not just making the art, they are making the very spaces in which the art happens' (Ibid.). Here, ecologically engaged practice is less 'a vision of self-expression and more about creating the conditions for change to occur' (Ariana Jordao in Allen et al. 2014, 22). Expanding further:

> There is a sense in which these artists put themselves in a position where they can be 'acted through'. The artist then becomes spokesperson, agitator, mediator, medium and, in some cases, healer…These artists tend to mix more with scientists, builders, economists, energy specialists, farmers and future scenario planners than with the peers they might have trained with…they are fluid and flexible and their activity either suggests adaptive changes that re-establish equilibrium or creates a new order…This demands a great deal of risk-taking and trust in imagination…an ability to withstand oppositional points of view, uncertainty and transitional states. (Allen et al. 2014, 27–282)

In the Howlround series on *Theatre in the Age of Climate Change,* Bilodeau (2015c) identified characteristics emerging in ecological performance making, particularly in response to climate change. They included an increased awareness of human-nature connection, stronger transdisciplinary engagement between artists and ecologically orientated disciplines, a growing interest in hybridity and non-conventional formats, a closer engagement with audiences through participatory initiatives, a willingness to embrace sustainability as a creative challenge and an effort to take 'work outside of traditional venues and into communities'. These characteristics are also synergistic with the 'social turn' (Bishop 2012) where 'artists work in forms of context-responsive and collaborative practice with communities, using a range of aesthetic, political and educational strategies to identify and affect local issues' (Badham 2013; Badham et al. 2020). Artists are now asking: 'How can we transform ecological concerns into compelling and environmentally-conscious theatrical experiences?'

The Agency and Potential of Scenography

Recent developments in the field of scenographic scholarship offer a platform from which to consider the role of the theatre designer for an ecological paradigm. Since the second half of the twentieth century, the term 'scenography' has rapidly expanded and now represents a progressive field that is moving far beyond traditional scenic illustration and naturalistic representation (Pitches and Popat 2011). As Arnold Aronson (2008, 23) writes, 'In recent years, scenographic and performative borders have shifted in ways so profound as to call into question the very notion of theatre and performance as it has been understood for over 2,500 years'. This means that 'opportunities for the designer have never been so varied or the territory so uncharted…we seem to be in a particularly acute stage of transition or collision in an indeterminate and shifting field' (Collins and Nisbet 2010, 1). No longer considered a mere backdrop for action, performance design has moved out of the shadows to not only confront but also blur traditional theatrical hierarchies of roles and functions. This blurring of boundaries is expanding our discipline to one that I suggest makes it more conducive to socio-ecological processes.

The most recent shift in scenographic discourse is arguably one that acknowledges scenography as an 'expanded disciplinary field' no longer bound by former traditions of theatre practice (Lotker 2015, 7). Under the notion of the 'expanded', narrative, performative and process-based strategies of scenography are re-explored to consider spatial politics and social agency (Brejzek 2010, 112). Here, 'the scenographer emerges not as the spatial organizer of scripted narratives but rather as the author of constructed situations and as an agent of interaction and communication' (2011, 14). The scenographer as world maker (Hann 2018) gives agency to the designer's capacity to transform literal and hypothetical worlds, creating distinct atmospheres that affect the way in which we view and engage with our environments. Contemporary performance designers are increasingly finding themselves straddling multiple disciplines and navigating diverse communities to seek possibilities for political, social, cultural and ecological revitalisation well beyond the confines of the theatre building (Irwin 2010, 2017; Lotker and Gough 2013;

McKinney and Palmer 2017; Hann 2018). Never before have we seen such a focus on the 'interactive relationships between buildings and bodies, between symbols of power and authority that structure public space and lived experience' (McKinney and Palmer 2017, 15).

Scenographer and scholar Kathleen Irwin (2017, 121) explains how designers in the realm of expanded scenography operate 'inside a burgeoning network of possibilities, interconnectivities and co-constituted intra-actions, working to make the most of the situation in relation to what is afforded by the circumstances'. Expanded scenography has opened up our discipline to not only start questioning 'conventional roles and sites of theatre' (Hannah 2015, 128) but also query long and tightly held assumptions about how we make theatre. This new-found porosity also makes space for theatre makers and designers within more traditional theatre venues and contexts to question conventional ways of doing things.

In the introduction to 'On Scenography' edition of *Performance Research* (2013, 6), editors Sodja Lotker and Richard Gough summarised 'scenographies as places, sites and locations where bodies and environments collide; scenographies as territories to be occupied, acted in and acted through but more than anything, territories where we have to take responsibility'. If the notion of the scenographer 'taking responsibility' is becoming an active phenomenon in contemporary practice, then perhaps it needs to be taken more seriously. Theatre makers and performance designers work across a diversity of platforms, from high-art cultural institutions and commercial sectors to grassroots initiatives. Our ability to simultaneously navigate a wide range of projects is at the crux of what makes us versatile, effective and employable as practitioners: 'theatre can serve as a kind of petri dish, a microcosm of the larger world where artists can test ideas, imagine new modes, and model change' (Goldmark 2019, 21).

While the expanded realm has opened up our sector to one which may be more accessible to integrating ecological ways of doing things, failing to navigate sustainability across a variety of platforms will reinforce its place only at the fringes of practice. Sustainability must be accessible to all practitioners, regardless of their workplace. As this ecological opportunity gains currency, the way in which the contemporary theatre maker navigates through this complex terrain also presents an exciting new challenge.

Proposing Ecoscenography

In *Performance and Ecology: What Can Theatre Do?* (2016, 229), Carl Lavery posits that 'if theatre is to contribute to a more progressive environmental future, then it is incumbent on practitioners and scholars to reflect, rigorously, on how theatre works as a medium, while at the same time remaining vigilant with respect to its supposed efficacy'. If, as Lavery posits, theatre has the 'capacity to alter how we exist in the world by troubling conventional modes of thinking and feeling', what does that mean for the way we make theatre? Could an ecological approach to performance practice entail what dramaturge Dillon Slagle (2013) refers to as 'an advancement in the craft of theatre creation akin to electric lighting, microphones, or the shift to realism'?

Beyond the necessity of energy and waste reduction, it is incumbent upon us to interrogate what an ecological approach to scenography *does*—how it influences our ways of thinking and working—as well as how it might be defined within and beyond the performing arts. In *Performing Nature* (2005), editors Gabriella Giannachi and Nigel Stewart (2005, 20) define ecology as 'an important model for cultural observation' explaining that it is the 'interface between ecology and the arts that some of the most aesthetically inspiring *and* politically challenging works are found' (emphasis in the original). However, in *Readings in Performance and Ecology* (2012) editors Wendy Arons and Theresa J. May (2012, 1) explain how 'performance and ecology do not easily or readily share space together, either materially or ontologically' and require that practitioners overcome preconceived binaries between performance and nature. While 'thinking about nature in performative terms can be challenging and disorientating', ecology opens up new forms of knowledge in which performance practice can be re-examined and re-invented (Szerszynski et al. 2003, 10).

In the *Performance Research* issue *On Ecology* (2012), Steve Bottoms, Aaron Franks, and Paula Kramer explain how ecological performance differs from notions of 'environmental theatre' and other site-based performance which ultimately views our surroundings (or the environment) as 'the scenic backdrop to an anthropocentric drama' in which 'definitions of environments refer to surroundings or external conditions, implicitly reaffirming humans as the centre of the conceptual equation' (Bottoms et al. 2012, 1). Instead, an ecological ethos is one that 'focuses on the development of a reciprocal, relational approach when working with people, place, context or material…the meaning and nature of which emerge through immersive, time and place-based processes' (Green 2015). Critical to differentiating the ecological ethos is viewing place as 'relational, part of a network, something connected' (Lavery and Whitehead 2012, 117). We should view place as the spatial manifestation of a broader system with which we wish to engage.

My premise in bringing an ecological approach to scenography—what I propose as 'Ecoscenography'—is to shift performance design to an increased awareness of broader ecologies and global issues: to conceptualise ways in which an ecological ethic can be incorporated into theatre practices (Beer 2015). Ecoscenography builds on contemporary reconsiderations of performance design, where creative and environmentally conscious processes align to become a fundamental part of the scenographer's ideas, practices and aesthetics. Being 'ecological' means integrating an awareness that no decision stands on its own: every design choice is intertwined with social, environmental, economic and political consequences that are far reaching and capable of having long-term effects and, ultimately, benefits.

Ecoscenography calls for a new approach to theatre production that overturns traditional production models. Like the revolution taking place in our societies and economies, this shift will not be easy. Yet, if issues such as climate change are 'driven by cultural values, logic suggests it can be tackled by shifting them' (Latham 2019). The key will be to see our current ecological crisis as a social, cultural, political and environmental opportunity. It is only by embracing the creative potential of sustainability that the performing arts will find its own path on the eco-agenda.

I propose that it is by moving beyond ecological narratives of despair that 'forays into a culture of repair, recuperation, and mutual value' (Svensen 2019) can occur. Ideas of 'hope', 'mutuality' and 'reparation' are at the crux of Ecoscenography, and ultimately at the transition to an ecological approach to theatre production.

The chapter that follows seeks to embed this ethos—this aspiration—in the broader concepts of ecological thinking. My aim is to demonstrate the theoretical foundations of Ecoscenography and outline the concepts—many of which are well established in related fields—that will enable eco-literate performance makers to put their intentions into practice. In short, what follows is an introduction to the 'thinking' that can power theatre's ecological transformation.

Notes

1. Examples include Julies Bicycle (UK), Creative Carbon Scotland (UK), Ecostage (UK), Sustainability in Production Alliance (UK), Green Arts Alliance (EU), Mitos21 (EU), Center for Sustainable Practice in the Arts (US/Canada), Greener Live Performances (Australia), Greenie in Residence (Australia), Howl Round Theatre in the Age of Climate Change series (USA), Theatre Green Book (UK), Ecosceno (Quebec, Canada) and Sustainable Performing Arts Now (Denmark).
2. Garrett took up this position at York University, Canada, in 2012. It is arguably still the only position of its kind in the world.
3. Notable events included Performance, Ecology and Responsibility symposium (Canterbury Christ Church University, UK); a colloquium on the Non-human and the Inhuman in Performing Arts (Theatre Academy Helsinki, Finland); UNFIX Festival of Performance and Ecology (Glasgow, UK); ART+CLIMATE=CHANGE (Melbourne, Australia); ArtCOP21 (worldwide); Performance Studies International (Performing Climates, Melbourne, 2016); Critical Theatre Ecologies (29th Annual CDE Conference, Augsburg, 2020).
4. Many of these artists have no doubt been influenced by the ecologically inspired performance artists of the 1960s, such as Joseph Beuys, Agnes Denes and Mierle Laderman Ukeles, as well as more recent work of artist/biologist Kathryn Miller, engineer/artist Natalie Jeremijenko and visual artists Lucy and Jorge Orta to name but a few.
5. Rather than following a 'traditional model' of success measured by flagship and high-profile venues, artists are increasingly 're-addressing or re-making' their own practices, letting go of former 'reputations or identities in favour of uncertainty and potential isolation' (Allen et al. 2014, 14).

References

Adebayo, Mojisola. 2009. What Would You Keep from Theatre and Performance Practice and What Needs to Change in Response to Climate Instability? In *The Ashton Directory*, ed. Wallace Heim. http://www.ashdendirectory.org.uk/featuresView.asp?pageIdentifier=2009521_19735354.

Allen, Paul, Emily Hinshelwood, Fern Smith, Rhodri Thomas, and Sarah Woods. 2014. *Emergence Culture Shift: How Artists Are Responding to Sustainability in Wales*. Cardiff, Wales: Art Council of Wales.

Arons, Wendy, and Theresa May. 2012. *Readings in Performance and Ecology*. New York: Palgrave Macmillan.

Aronson, Arnold. 2008. The Power of Space in a Virtual World. In *Performance Design*, ed. Dorita Hannah and Olav Harsløv. Copenhagen, Denmark: Museum Tusculaneum Press.

Badham, Marnie. 2013. The Turn to Community: Exploring the Political and Relational in the Arts. *Journal Of Arts & Communities* 5 (2/3): 93–104.

Badham, Marnie, Kitt Wise, and Abbey MacDonald. 2020. Mona's 24 Carrot Gardens: Seeding an Ecology of Cultural Value. In *Exploring Cultural Value: Contemporary Issues for Theory and Practice*, ed. Kim Lehman, Ian Fillis, and Mark Wickham. Bingley, UK: Emerald.

Barnard, Dan, and Rachel Briscoe. 2015. The One About the Tree Falling in the Forest: A Practice-Based Reflection on the Stories We Tell and Who We Tell Them To. Paper presented at the Performance, Ecology and Research Symposium, University of Christchurch, Canterbury, New Zealand.

Beer, Tanja. 2015. The Living Stage: A Case Study in Ecoscenography. *Etudes* 1 (1): 1–16.

Beer, Tanja, Lanxing Fu, and Cristina Hernández-Santín. 2018. Scenographer as Placemaker: Co-creating Communities Through The Living Stage NYC. *Theatre and Performance Design* 4 (4): 342–363.

Bilodeau, Chantal. 2015a. Nurturing Local Seeds into Global Vibrancy: Climate Change Theatre Action. In *Howlround Theatre Commons*. https://howlround.com/nurturing-local-seeds-global-vibrancy.

———. 2015b. Our Affair With Energy. In *Artists and Climate Change: Building Earth Connections*. https://artistsandclimatechange.com/2015/06/05/our-affair-with-energy/.

———. 2015c. In Search of a New Aesthetic. In *Howlround Theatre Commons*. https://howlround.com/search-new-aesthetic.

———. 2019. What I Learned About Gender Parity and Racial Diversity from Running a Global Participatory Initiative. In *Howlround Theatre Commons*. https://howlround.com/what-i-learned-about-gender-parity-and-racial-diversity-running-global-participatory-initiative?fbclid=IwAR3U3ezfJbxMfHPVLkksXf9YsVN6XFyqOHQYo-IrbQ6e0uP-JqTk1qHYqcs.

Bishop, Claire. 2012. *Artificial Hells: Participatory Art and the Politics of Spectatorship*. London, UK: Verso Books.

Bottoms, Steven, Aaron Franks, and Paula Kramer. 2012. On Ecology. *Performance Research* 17 (4): 1–4.

Brejzek, Thea. 2010. From Social Network to Urban Intervention: On the Scenographies of Flash Mobs and Urban Swarms. *International Journal of Performance Arts & Digital Media* 6 (1): 109–122.

———. 2011. Space and Desire. In *Space and Desire: Scenographic Strategies in Theatre, Art and Media*, ed. Thea Brejzek, Wolfgang Greisenegger, and Lawrence Wallen. Zurich, Germany: ZHdK.

Brunner, Paul, and Michael Mehler. 2013. Theatre Design and Production Reimagined: Four Principles for a Sustainable Future. *Theatre Design and Technology* 49 (3): 23–32.

Brunner, Paul, and Olivia Ranseen. 2017. The Greening of Academic Theatre: While Several Impediments Hinder Successful Long-Term Integration of Sustainable Practices, They Aren't Insurmountable. *TD&T (Theatre Design & Technology)* 53 (3): 20-33.

Butler, Robert. 2009. Green Shoots of Climate-Change Theatre. *The Guardian*. https://www.theguardian.com/stage/theatreblog/2009/may/22/climate-change-theatre.

Cattenazzi, Adele. 2019. *The Life-Cycle of a Set Piece: Exploring Post-Production Waste in Set Design in Melbourne*. Australia: Monash University.

Collins, Jane, and Andrew Nisbet. 2010. *Theatre and Performance Design: A Reader in Scenography*. London, UK: Routledge.

Edensor, Tim. 2005. *Industrial Ruins: Spaces, Aesthetics and Materiality*. Oxford, UK: Berg Publishers.

Fried, Larry, and Theresa May. 1992. *Greening Up Our Houses: A Guidebook to an Ecologically Sensitive Theatre Organization*. New York: Drama Book Publishers.

Garrett, Ian. 2012. Theatrical Production's Carbon Footprint. In *Readings in Performance and Ecology*, ed. Wendy Arons and Theresa May, 201–209. New York: Springer.

Giannachi, Gabriella, and Nigel Stewart. 2005. *Performing Nature: Explorations in Ecology and the Arts*. Hochfeldstrasse, Switzerland: Peter Lang.

Goldmark, Sandra. 2017. Theatre 'Stuff:' Use and Reuse in the Theatre: Examining Our Human Entanglement with Stuff and Mitigating Its Environmental Impact. *TD&T (Theatre Design & Technology)* 53 (1): 20–29.

———. 2019. Circular Design and Production. *TD&T (Theatre Design & Technology)* 55 (1): 10–21.

Green, Nic. 2015. How to Love the Raindrops on Your Washing Line: Towards an Ecological Performance Making Practice. In *Create & Sustain*. GSA of Sustainability, Glasgow School of Art, UK. http://www.gsasustainability.org.uk/archive/create-sustain-nic-green.

Guggenheim, Davis (Director). 2006. *An Inconvenient Truth: A Global Warning*. Accessed November 24, 2021. https://www.amazon.com/Inconvenient-Truth-Al-Gore/dp/B000KZ3BWE.

Hann, Rachel. 2018. *Beyond Scenography*. London, UK: Routledge.

Hannah, Dorita. 2015. Constructing Barricades and Creating Borderline Events. *Theatre & Performance Design* 1 (1–2): 126–143.

Heim, Wallace. 2013. Investigating the Aesthetics of Ecological Design and Eco-Scenography. In *World Stage Design: People, Planet and Profit*. Cardiff, Wales. http://wallaceheim.com/investigating-the-aesthetics-of-ecological-design-and-eco-scenography/.

Heinberg, Richard, and Daniel Lerch. 2010. *The Post Carbon Reader: Managing the 21st Century's Sustainability Crises*. California, USA: Watershed Media Press.

Howard, Pamela. 2009. *What is Scenography?* Oxfordshire, UK: Routledge.

Irwin, Kathleen. 2010. Canada: Remembering and Un-remembering a Century of Prairie Settlement – Community-Triggered Performance. *Local-Global: Identity, Security, Community* 7: 120–133.

———. 2017. Scenographic Agency: A Showing-Doing and a Responsibility for Showingdoing. In *Scenography Expanded: An Introduction to Contemporary Performance Design*, ed. Joslin McKinney and Scott Palmer, 111–124. London, UK: Bloomsbury Methuen Drama.

Klein, Naomi. 2014. *This Changes Everything: Capitalism vs. the Climate*. New York: Simon & Schuster Paperbacks.

Latham, Lucy. 2019. The Creative Climate Movement. In *Howlround Theatre Commons*. https://howlround.com/creative-climate-movement.

Lavery, Carl. 2016. *Performance and Ecology: What Can Theatre Do?* Oxfordshire, UK: Taylor and Francis.

Lavery, Carl, and Simon Whitehead. 2012. Bringing It All Back Home: Towards an Ecology of Place. *Performance Research: On Ecology* 17 (4): 111–119.

Lawler, Mike. 2008. Toward a More Sustainable Theatre: Why Theatre Artists Should Be Concerned About Pollution, Global Climate Change and the Sustainable Future. *American Theatre* 25 (7): 58.

Leonardi, Thierry. 2018. OSCaR – A Project for Opera Sceneries Circularity and Resource Efficiency. In *Economie Circulaire*. https://www.economiecirculaire.org/community/pg/groups/6832/.

Lotker, Sodja. 2015. Expanding Scenography: Notes on the Curatorial Developments of the Prague Quadrennial. *Theatre & Performance Design* 1 (1–2): 7–16.

Lotker, Sodja, and Richard Gough. 2013. On Scenography: Editorial. *Performance Research: A Journal of the Performing Arts* 18 (3): 3–6.

McDonough, William, and Michael Braungart. 2002. *Cradle to Cradle: Remaking the Way We Make Things*. New York: North Point Press.

McKibben, Bill. 2005. *What the Warming World Needs Now Is Art, Sweet Art*. Accessed November 24, 2021. https://grist.org/article/mckibben-imagine/.

McKinney, Joslin, and Scott Palmer. 2017. *Scenography Expanded: An Introduction to Contemporary Performance Design*. London, UK: Bloomsbury Methuen Drama.

Morris, Damond. 2007. *Towards a Recycled Theatre: Industrial Ecology Theatrical Applications for the Next Industrial Revolution*. Bellingham, WA: Western Washington University.

Omen, Katy. 2014. Stage Lighting and the Environment. *Lighting & Sound America*. http://www.lightingandsoundamerica.com/reprint/EnvironmentStageLight.pdf

Peeters, Jeroen. 2012. Imagination, Experience and Meaning as Quality of Life: The Performing Arts and Sustainable Development in Flanders. In *Ins & Outs: A Field Analysis of the Performing Arts in Flanders*, ed. Joris Janssens. Brussels, Belgium: Vlaams Theater Instituut.

Pitches, Jonathan, and Sita Popat. 2011. *Performance Perspectives: A Critical Introduction*. New York: Palgrave Macmillan.

Slagle, Dillon. 2013. The Aesthetic Evolution of Eco Theater. In *Howlround Theatre Commons*. https://howlround.com/aesthetic-evolution-eco-theater.

Stancato, Jon. 2010. Sustainable Sustenance: Launching Community Supported Theatre. *Centre for Sustainable Practice in the Arts Quarterly* 1 (2): 36.

Svensen, Zoe. 2019. The Factory of the Future: A Collaborative Project for Imagining Otherwise. In *Howlround Theatre Commons*. https://howlround.com/factory-future.

Szerszynski, Bronislaw, Wallace Heim, and Claire Waterton. 2003. *Nature Performed: Environment, Culture and Performance*. Oxford, UK: Blackwell.

2

Ecological Thinking

T. Beer, *Ecoscenography*, https://doi.org/10.1007/978-981-16-7178-4_2

We live in a system that has increasingly favoured humanistic values over earth-centred ethics. A theatre maker's attempts to work ecologically today are hindered by anthropocentric worldviews and power relations that began centuries ago. As theatre composer François Ribac (2018, 52) contends:

> [Our] transition to fossil fuels at the end of the eighteenth century was not part of a natural progression, but rather must be traced back to the development of capitalism in Europe and to forms of domination and privatization of resources, land and bodies.

Ecoscenography's focus on relational thinking—one where social and environmental justice is intertwined with artistic pursuits—entails confronting long-held ontologies and belief systems that have consciously or unconsciously influenced our ways of perceiving, thinking and ultimately working in an ecological way.

Since the emergence of the Scientific Revolution in the 1500s, and particularly the Industrial Revolution of the 1800s, Western civilisation has continued to believe in a system that separates humans from the ecological systems from which we are a part. A number of commentators (e.g. Schumacher 1973; Rees 1995; Capra 1996; Hawken et al. 1999) have increasingly drawn attention to this dominant anthropocentric worldview, which has encouraged us to see ourselves as separate from nature—to view the world as something to be analysed and exploited with little regard for environmental consequences. Ecofeminist philosopher and historian Carolyn Merchant (1980) explains how it is largely this separated thinking and reductionist view of the world that has propagated a disconnection from nature, leaving us unbalanced in our understanding of ecology and the living world. This perspective has also dominated discourses of sustainability across multiple fields, largely distancing and disconnecting us from the problems at hand, whilst limiting the efficacy of our engagement with the living world (Bateson 1972; Clayton and Radcliffe 1996).

This chapter is concerned with the thinking that underpins my conceptualisation of Ecoscenography. It brings together multiple philosophies and perspectives across fields of environmental science, deep ecology, ecofeminism, ecomaterialism and Indigenous knowledge systems to lay the foundations of 'ecological thinking'—an umbrella term that I use to draw attention to the extricable connections (or intra-activity) between living systems and human-centred design. Ecofeminist author Lorraine Code (2006, 25) highlights how ecology 'literally and metaphorically, affords a model for rethinking the established theories of knowledge, and relations between humanity and the other-than-human'. While scientific notions of ecology focus on understanding nature in terms of processes and relationships that produce ecosystems, ecological thinking is concerned with understanding the flows, relationships, interdependence, interconnection, co-existence, co-evolution and consciousness that make up our living world in its entirety, humans included (Capra 1996; du Plessis and Cole 2011; Hes and du Plessis 2014; du Plessis and Brandon 2015).

Ecological thinking is a departure from the social fantasy of the post-Enlightenment Western world that advocated for man's hierarchical dominance over nature to instead take up a holistic and reciprocal positioning with the more-than-human world (Code 2006). It 'conjures up images of something that is ever-changing, in motion, and not static' (Allen 2019, 121). To think *ecologically* is to be 'attentive to the capacity of relation-creation, to how different beings affect each other, to what they do to each other, the internal "poiesis" of a particular configuration' (Puig de la Bellacasa 2016, 52). This 'attentiveness' to relational systems, to the 'affectivity' of our practices, is at the heart of bringing an ecological ethic into performance design.

Towards an Ecological Worldview

It has been extensively argued that for society to move towards a more ecological ethos, we need to change the existing paradigm in which we operate (McHarg 1969; Goldsmith 1992; Rees 1995; Capra 1996; Hawken et al. 1999). Current understandings of our ecological crisis are dominated by linear, rationalist and reductionist perspectives that propagate notions of polarisation, 'scarcity thinking', 'fear-based narratives', 'fragmentation' and 'disconnection' (Hes and du Plessis 2014, 12–13).

This mechanistic worldview tends to see nature as a machine-like system, where the properties of the whole could be reduced to the sum of its parts—as is possible with mechanical and other complex cybernetic systems, such as factories, power grids and computers (Abram 1997; Drengson 1995; Riskin 2015). From a mechanistic worldview, humans are masterful owners of nature, where society's activities and its impacts can be measured and controlled. This individualistic 'machine-like' or technocratic focus has led to a psychological separation of humans and living systems, largely limiting the efficacy of the sustainability movement and, ultimately, producing the current ecological crisis we find ourselves in today.

While the mechanistic worldview has provided valuable scientific and quantitative insights into how the world functions, it does not acknowledge 'all of reality', including the full potential of the global social-ecological system of which we are a part (Hes and du Plessis 2014, 23). To account for this discrepancy, an alternative perspective based on the study of living systems—summarised under the notion of 'the ecological worldview'—has emerged from fields as diverse as ecology, quantum physics, systems thinking, neuroscience, Indigenous studies, psychology and sociology (e.g. McHarg 1969; Goldsmith 1992; Capra 1996). In contrast to dominant discourses of sustainability that favour dispassionate, linear and quantitative methodologies or measurement processes, the ecological worldview holds that the properties (and behaviour) of the whole are more than the sum of, and not deducible from, the properties of the parts (du Plessis and Cole 2011, 438). The ecological worldview also 'interrogates and endeavors to unsettle the self-certainties of western capitalism and the epistemologies of mastery it underwrites' (Code 2006, 4).

The ecological worldview incorporates principles of wholeness, interdependence, diversity, partnership, flows, flexibility and cycles (DeKay 2011, 65). In *Designing for Hope: Pathways to Regenerative Sustainability* (2014, 159), sustainability experts Dominique Hes and Chrisna du Plessis identify three common values underpinning this phenomenon. These are (1) 'wholeness'—a shift in emphasis from a focus on the 'parts' to the 'whole', leading to values such as inclusiveness and interconnectedness; (2) 'relationship'—a relational view of positive and mutually supportive relationships, leading to values of mutuality, fellowship, positive reciprocity, responsibility, respect and harmony; (3) and 'change'—accepting the inherent complexity, impermanence and unpredictability of the world, leading to values of humility and non-attachment. These values underpin the fundamental framework that constitutes a more ecologically engaged practice.

The ideologies of the ecological worldview entail that humans consider themselves not as separate from nature, but as inextricably embedded in the 'web of life' (Capra 1996). As philosopher Arne Næss argues in *Ecology, Community and Lifestyle* (1989, 79), 'a human is not a thing…but a juncture in a relational system without determined boundaries in space and time'. Living systems are relational systems that connect humans to abiotic and biotic structures to acknowledge ourselves as 'co-creators in the co-evolution that shapes the living world' (Hes and du Plessis 2014, 39). Thus, moving towards an ecological worldview requires that we find ways to reconsider our practice outside of the dominance of 'mastery rationalism and humanism' that governs much of contemporary society (St. Pierre 2019, 102).

Ancient and Indigenous communities have long advocated for inextricable connections between humans and other living systems. In *Sand Talk: How Indigenous Thinking Can Save the World* (2019), Tyson Yunkaporta (Apalech Clan) writes how Indigenous notions of kinship, reciprocity and more-than-human belonging are essential elements of moving towards a thriving future. In *Nourishing Terrains* (1996, 8), Deborah Bird Rose highlights how Indigenous Australian perspectives include an understanding of the world through the notion of 'Country', a multi-dimensional concept and living entity which includes 'people, animals, plants, Dreamings, underground, earth, soils, minerals and waters, surface water, and air'. These ideas go far beyond physical elements of environmental conditions to one that involves a reciprocal relationship of identity that includes cultural knowledge. For example, 'Connection to Country' is a fundamental notion of kinship in which Indigenous Australian perceptions of land, health, spirit, story and culture are an inseparable part of Australia's First Nations identity. As Rose writes:

> People talk about country in the same way that they would talk about a person: they speak to country, sing to country, visit country, worry about country, feel sorry for country, and long for country. People say that country knows, hears, smells, takes notice, takes care, is sorry or happy…for many Aboriginal people, everything in the world is alive: animals, trees, rains, sun, moon, some rocks and hills, and people are all conscious…the concept of a living world depends on communication: it requires that one listen as well as speak; it requires an attitude of attentiveness and a degree of respect. I would say that the concept of a truly living world requires a shift in thinking for many settlers (1996, 7, 23, 85).

Rose highlights how the act of listening, of respectful communion with the more-than-human world, is integral to engaging in a reciprocal dialogue with our land, our communities, our kin, our co-extensive selves.

Indeed, Aboriginal Australian conceptualisations of environmental custodianship provide an important source of inspiration and guidance when considering how we co-create cultural practices that are in tune with living systems. Current estimates suggest an immense antiquity to Indigenous Australian culture, in the order of more than 60,000 years—the longest continuous culture in the world (e.g. Calma 2010). Indigenous Australians show us how 'performing' social-ecological reparation and connection is vital for human societies to survive through deep time. Australia's First Peoples have demonstrated how the ecological is cultural and that the merging of art, land, science and story must be part of our collective healing and revitalisation. As many Indigenous Australian thinkers have highlighted (e.g. Pascoe 2014), painting, weaving, stories, songs and dances are fundamental to knowing, healing and contributing to Country. While First Nations cultures are incredibly diverse and unique, these congruent ways of seeing and knowing inspire a stronger relationship with ourselves and our environment.

Adopting an ecological worldview has to do with how we imagine ourselves as part of nature (Morton 2010) and 'the way living entities, including humans, relate to, interact with and depend upon each other' (Mang and Reed 2015, 8). This involves altering the lens through which we perceive the world and ourselves (Kegan 1982) as well as a broadening of identity in how we acknowledge ourselves in relationship to the world around us. Hes and du Plessis (2014, 156) explain that 'as one's identity expands, so does one's view of the world…With these changed perceptions also come a change in values, behaviours and possible leverage points'. At the core of this ideological shift is a change in focus, a moving away from anthropocentric or mechanistic thought (separateness) to include concepts of integration, awareness and holistic perception (interconnectedness). Integral theorist Sean Esbjörn-Hargens (2009) describes this 'widening of identity' as a transition from 'me' (egocentric) to 'my group' (ethnocentric) to 'my country' (sociocentric) to 'all of us' (worldcentric) to 'all beings' (planetcentric) to finally 'all of reality' (kosmoscentric). The idea that 'the extended awareness of the self' is 'part of an interconnected whole that is also part of the self' draws our attention to 'the idea that we are in this together, and what happens to the "other" will also have an effect on the self' (Hes and du Plessis 2014, 35).

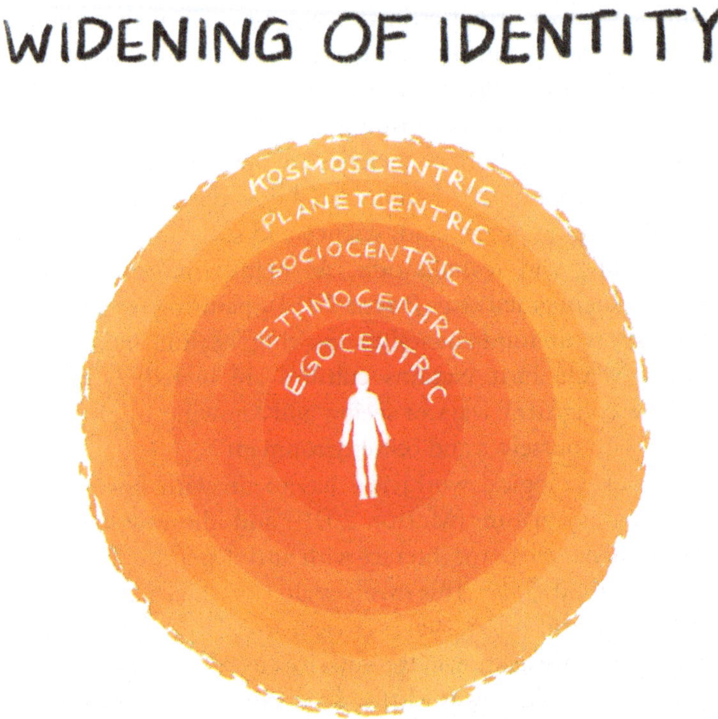

In Ecoscenography, I interpret this shift for the theatre artist as a kind of 'creative expansion'—a widening in identity that expands from the individual practitioner, to acknowledging the multiple interconnections across communities and environments in the making of the work that we do. To bring an ecological worldview into theatre production requires that we question some of our underlining assumptions and attitudes regarding sustainability to understand that the opportunities of ecological thinking emerge through the action of participating in and co-evolving with living systems. This transition is not necessarily seen as oppositional to more mechanistic or technocratic ways of doing things. For example, while quantitative tools such Julie's Bicycle's 'IG tool' carbon calculator can provide essential information for tracking material and energy usage in theatre production (Jones 2014, 25), an ecological perspective also takes into account relational knowledge and experiences—the unique context of the theatre venue and wider community—to make informed decisions. This merging of worldviews makes space for the creative capacity of ecological themes and processes. In Ecoscenography, the 'sustainability credentials' of a theatre production focus not only on energy and waste reduction (a primary concern of the mechanistic worldview), but also on the project's capability to connect audiences to creative and ecological processes through the sharing of ideas and resources. This expanded and integrative approach to sustainable practice will be discussed further in Chaps. 3 and 4.

In summary, shifting to an ecological worldview in theatre production demands a new way of thinking, of how we make work that is more conscious and connected to the environmental, social and cultural systems of which we are part. Simply put, Ecoscenography challenges our modern and prevailing individualist focus (a mindset that favours the 'ego') to consider the wellbeing of the whole of which we are also part (one that values the 'eco'). Integral theorist Mark DeKay (2011, 60) explains that this is no easy cognitive task, but rather part of a transition in our developing capacity as humans. Despite this challenge, I suggest that this transition offers opportunities for theatre makers to think beyond the transient qualities of the theatre or site, to also understand how their work affects wider communities and living processes.

Ecomaterialism: Embracing the More-than-human

Ecoscenography is inspired by new ways of conceptualising more-than-human relations. While notions of the ecological worldview have helped reframe sustainability beyond its dominant mechanistic framework, a proliferation of studies under the banner of 'new materialism'[1] has also called for new preoccupation with the interconnectedness of the living world. This 'material turn'[2] has caused a radical shift in environmental humanities that embodies twenty-first-century agendas implicit in the ecological crisis (Iovino and Oppermann 2012a, 75). The ecological focus of new materialism—ecomaterialism—argues for the dissolving of the subject-object binary of the material world, incorporating notions of agency as a way of generating awareness of ecological issues (Cohen and Duckert 2013, 4). Its theories collapse human and non-human dichotomies, realigning the human with 'a world of biological creatures, ecosystems, and xenobiotics, humanly made substances' in which the material self becomes entangled within a web of political, ecological, economic and climatic forces (Alaimo 2010a, 115). Essentially, ecomaterialism 'defies reductionist, dualistic modes of opposition and hierarchies ("binary thinking") which posits humans as oppositional to non-humans in favour of connecting with more complex, egalitarian and ecological modes of being' (Beer 2016a, 162).

Much like the ecological worldview, ecomaterialism augments the liveness of the planet by unifying the human self with the broader system of ecological organisation. Ecomaterialism explicitly abandons the 'nature' and 'culture' divide to re-examine material, biological and social constructs and relationships. While the mechanistic stance of sustainability 'offers a comforting sense of scientific distancing and objectivity' (Alaimo 2012, 561, 563), ecomaterialism calls for the merging of human and non-human agencies (including their linguistic, social, political and cultural representations) to stimulate a better understanding of environmental issues (Iovino and Oppermann 2012a, 84). This includes exploring how the relational dynamics between humans and non-humans have interrogated more sustainable modes of production 'in the name of vigorous materiality rather than in the name of the environment' (Bennett 2010, 111). Questioning the separation of the material self from the 'more-than-human' world, ecomaterialism asks: 'Can sustainability be transformed in such a way as to cultivate posthumanist[3] epistemologies, ethics, politics, and aesthetics?' (Alaimo 2012, 561, 563).

At the heart of ecomaterialist ontologies is feminist thinker and quantum physicist Karen Barad's concept of 'intra-action', which involves moving away from binaries that separate matter, agency, human, non-human, nature and culture, to embrace the complex intra-relationships (intra-action) between all things. Extending beyond 'interaction' (that which occurs between distinct entities), Barad's term, 'intra-action' refers to a fundamental and co-creative entanglement between individual entities (Barad 2007, 128). For Barad, nothing exists as a thing-in-itself but is 'worlding in its materiality' in which matter is not 'fixed' in reality but is continuously evolving and reacting to the world in motion (181). 'Worlding' relates to how all things are interconnected, temporal, dynamic, affective and causational. Entities—humans, plants, theatres, LED lights, weather, landscapes, bacteria, ocean plastic and landfills—exist in the phenomena of multiple relations as they co-create 'differential patterns of mattering' (181). Put more simply, matter is seen as an active agent in the intra-activity of the 'doing' or 'being' in the world (178). Similar to ideas of systems thinking, the concept of matter as 'intra-active becoming' (Barad 2003, 822) focuses on the fluidity of interfaces, interchanges and transformation, resisting 'ideological forces of disconnection' (Alaimo 2010a, 142) in favour of synthesising connections with the more-than-human world.

The idea that matter possesses agency is of central significance in ecomaterialism (e.g. Barad 2007; Alaimo 2010a; Bennett 2010; Frost and Coole 2010; Iovino and Oppermann 2012a). First acknowledged in Ilya Prigogine and Isabelle Stenger's *Order Out of Chaos* (1985, 9), early notions of the material turn led to a new view of matter which was 'no longer the passive inert substance described in the mechanistic world view' but a 'new dialogue' between humans and nature. In this context, nature and its agentic capacity are 'reanimated' and 'recognised' (Iovino and Oppermann 2012b, 464). All non-human things—water, soil, stones, metals, minerals, bacteria, toxins, electricity, cells, parasites and garbage—are viewed as vital, vibrant and alive, and therefore possessing 'agentic capacity' (461). Here, matter's agency or 'inherent creativity' (DeLanda 1997, 16) is acknowledged as 'the very substance of ourselves and others' (Alaimo 2010a, 158).

What is perhaps most distinct from other ontologies of ecological thinking is ecomaterialism's focus on 'both organic and inorganic matter' (Iovino and Oppermann 2012b, 464) where the concept of 'matter as alive' draws attention to both biological and technological 'intertwining' (Pickering 1995, 15). By absolving matter from its long history of mechanistic, static and passive attachment (Bennett 2010), the material turn considers how humans and all 'things' (both biotic and abiotic) might co-emerge into an integrated field of 'matter' existence. Although ecomaterialist arguments have been criticised by some scholars for being too 'homogenising' (e.g. Ingold 2012) and 'primitive'[4] (e.g. Rosenberg 2014), I suggest that they offer a way in which to resist nature/culture/human binaries in scenographic practice. As ecofeminist scholar Stacy Alaimo contends, breaking subject-object dichotomies is one way in which culture and environment can no longer be considered separate as 'there is, ultimately, no firm divide between mind and matter, organism and environment, self and world' (Alaimo 2011, 282, 283).

For political theorist Jane Bennett, it is by dissolving the subject-object binary of the material world that we can 'begin to experience the relationship between persons and other materialities more horizontally' thereby improving our 'ecological sensibility' (Bennett 2010, 10). A key text of the material turn, Bennett's *Vibrant Matter: A Political Ecology of Things* (2010) redefines the natural world outside the constraints of the mechanistic worldview and views the concept of matter as vibrant and forceful—an actant[5] in human experience. Considering ontological ideas about the relationship between humans and 'things', Bennett's 'vital materiality', 'thing-power' and 'distributive agency' incorporate notions of agency as a way of generating awareness of sustainability (40, 46, 21).

By rethinking our tendency to separate the world into 'dull matter (it, things) and vibrant life (us, beings)', Bennett reimagines things as actants and assemblages[6] which belong to the compound nature of the human self (111). For Bennett, 'vibrant materiality' (xenobiotic agents, food, minerals, parasites, dirt, waste, bodies) 'runs alongside and inside humans' (viii). It is in this context that Bennett explains how materiality can 'horizontalize the relations between humans, biota, and abiota … toward a greater appreciation of the complex entanglements of humans and nonhumans' (112).

Another key aspect of ecomaterialism is a shift 'from a sense of place to a…more systemic sense of planet' (Heise 2008, 56), which also coincides with the ecological worldview's broadening of identity. In *Bodily Natures* (2010a, 9) Alaimo's concept of 'trans-corporeality' uses this 'planet focus' to theorise 'material interconnections of human corporeality with the more-than-human world'. Focusing on invisible material forces—or 'flows of substances between people, places, and economic/political systems'—Alaimo tracks 'the often unpredictable and unwanted actions of human bodies, non-human creatures, ecological systems, chemical agents, and other actors' (2). Trans-corporeality accounts 'for the ways in which nature, the environment, and the material world itself signify, act upon, or otherwise affect human bodies, knowledge, and practices' (7–8). As Alaimo contends, 'thinking with complex mappings of agencies and interactions' the 'material self…cannot be disentangled from networks that are simultaneously economic, political, cultural, scientific, and substantial' (20).

The fact that trans-corporeality recognises material interchanges between local and global, bodies, substances, flows and forces have specific relevance to the potentially harmful substances found in scenographic practice. Alaimo's notion of 'toxic bodies' highlights the hazardously entwined social, ecological, political, cultural, and material forces that are present in 'trans-corporeal space' (Alaimo 2010b, 31). Moreover, the emphasis on toxicity and the body suggests how 'toxic bodies' may provoke ethics that move from 'disembodied values and ideals' towards acknowledging 'practices that have far-reaching and often unforeseen consequences for multiple peoples, species, and ecologies' (Alaimo 2010a, 283).

Integrating an ecological approach to theatre production requires that we consider the interconnectedness of not only biotic systems (humans, air quality, rivers) but also the abiotic components (plastics, air conditioners, synthetic dyes) that make up the more-than-human world. Ecomaterialism offers many ideas that may stimulate a way of thinking ecologically in scenographic practice, including a focus on materiality and bodily experience as a way of contextualising the co-extensive relationship of the human self with the environment (Alaimo 2010a). As Serenella Iovino and Serpil Oppermann (2012b, 450, 469) summarise, 'what is at stake in the "material turn" is the search for new conceptual models apt to theorize the connections between matter and agency on the one side and the intertwining of bodies, natures, and meanings on the other side…that co-constitute our existence'. For Alaimo (2012, 563–564), this means holding 'ourselves accountable to a materiality that is never merely an external, blank, or inert space but the active, emergent substance of ourselves and others'.

Ecoscenography draws inspiration from ecomaterialism through its focus on the complex entanglement of material processes, interactions and effects (both seen and unseen) as part of its processes. The material turn's focus on 'intra-active becoming' (rather than 'being') offers a chance for theatre makers to cogitate the potential of materials in and beyond the production, including a consideration of socio-political and environmental impacts. In addition, the notion of matter as active, transformative, self-organising and vibrant draws attention to the possible consequences of scenographic material and substances in and beyond theatre production. It emphasises the impacts of seemingly inert or invisible substances of 'agency' (e.g. unrecyclable set elements, highly toxic flame retardants and spray paints or cheaply mass-produced artefacts) due to a greater awareness of 'trans-corporeal' consequences such as contributing to landfill waste, air pollution and child labour. Additionally, the idea that materiality is emergent and co-extensive may bolster ecological considerations in the making, presentation and distribution of scenographic elements. Lastly, letting go of nature-human binaries may inspire designers to see themselves as part of the environmental effects (both positive and negative) that they are creating.

At the heart of ecological thinking is the framing of our inextricable interconnectedness with the more-than-human world, including a widening of identity—of what it means to be 'human'—in an age of environmental fragility. The above philosophies share an array of ontological views, such as a frustration with anthropocentric, fragmented and reductionist perspectives; a sense that nature is not 'outside', 'out there' or 'separate' from one's self; a realisation that the world is substantially interconnected, where nature and culture are intrinsically linked; and an acknowledgement of different levels of existence (physical, intellectual, emotional, social and spiritual) that solicits an active co-operative engagement with the more-than-human world. Together, they form a crucial part of Ecoscenography's theoretical framework.

Ecological Thinking in Theatre and Performance Design

Understanding how theatre scholarship and production may be understood through the lens of ecology, agency and corporeality has been a topic of considerable interest in recent years. For performance and ecology scholar Lisa Woynarski (2015, 32), ecological ontologies reject 'reductive binary-making practices' to 'open up new ways of thinking about "humanness" and "performance"'. The amalgamation of performance and ecology seeks to integrate itself with living processes in both philosophy and praxis—'to enact ecological thinking' in performance that 'can mediate our understanding of the way in which we are immersed and entangled with more-than-humans in an ecological sense and the way in which our perspectives on that immersion is shaped by social and cultural formations and representations' (Woynarski 2020, 13, 126). Further, Woynarski's concept of 'intersectional ecologies' views ecological thinking as one which also overlaps with socio-ecological 'injustices, exclusions and oppressions' to consider 'who is affected and marginalised, and whose voice or perspective is being heard and whose is being erased' (6). As Woynarski describes:

> Intersectional ecological thinking is both a way of looking and praxis. It is based on the idea that, on a global scale, ecological effects are unevenly disrupted and tied to social structures that disproportionately affect marginalised people such as women, people of colour, Indigenous peoples and the poor. (19)

While the field of performance and ecology is still emerging (see Heddon and Mackey 2012; Bottoms et al. 2012; Arons and May 2012; Allen and Preece 2016; Woynarski 2020), and requires more diverse viewpoints, at the crux of its shared epistemology is the notion that all theatre is connected with the ecomaterial world, affecting it and being affected by it. Bringing ecological thinking into the way in which we enact performance helps us 'to imagine, disrupt and/or reconcile climate-changed futures' (Woynarski et al. 2020, 185). It 'offers particular kinds of experience and knowledge about human relations with nature, with ecologies' (Heim 2014).

In her book, *Ecodramaturgies: Theatre, Performance and Climate Change* (2020), Woynarski states how theatre making offers an opportunity to critique or reveal the materiality of ecological interrelationships—to rethink our perceived separation between humans and the ecological world. For many practitioners, ecological thinking offers 'new actants for the stage' (Arons 2012, 571), as well as ways to 'enact and reveal the human–nonhuman assemblages and interconnections that make up the ecologically material world' (Woynarski 2015, 31). For example, UK dancer Paula Kramer (2012, 91) explains how ecomaterialist perspectives suggest 'an overall blossoming of all kinds of materials' emphasising the 'human body in the natural environment…as vibrant matter among vibrant matter, as lively thing among lively things'. Drawing from Jane Bennett's thesis (above), Kramer describes her own 'vital materialist' performative experience as 'an embodied and receptive engagement with a world full of thing-power' which involves 'continuous learning' to keep the 'mind in the body yet letting awareness travel far and wide'(Ibid.). This means approaching performance from a perspective where neither human nor material takes precedence and allowing both to inform and to communicate with each other (83).

Extending partnerships to give agency to non-human matter is now an established topic in scenography (McKinney and Palmer 2017). The material turn offers a unique perspective in the way theatre makers approach their practice, their choices of material and creative processes, bringing with it an increased interest in material sensitivities, hands-on making and more-than-human co-creative interaction. For example, scenography scholar Joslin McKinney (2015b) draws on Bennett's vital materialism to explore materiality and agency in scenography, including how theatre's traditional human and non-human dichotomies and hierarchies might be reconsidered. Using her own practice as precedent, McKinney reflects on ways in which scenographers, performers and audiences might explore the 'textility' and 'vibrancy' of their material encounters. Here, theatre materials take on a new 'life' where the focus of these substances is in 'their interaction rather than in their singular entities' (135). From an ecomaterialist perspective, it is the agentic capacities of materials that take priority in the scenographic experience, rather than what is imposed on them by the human.

Collaborating with material was a focus of scenographer Illka Louw's masters' thesis, *From Designer Through Space to Spectator: Tracking an Imaginative Exchange Between the Actants of a Scenographic Event* (2013), which explored the 'fluidity of the division between human bodies, bodies of space and bodies of materials' (3). Using vital materialism as context, Louw explored the relationship between human and non-human actants by placing herself as co-performer with a piece of fabric (12). Louw highlighted 'the blurred division between material, object, space and time' as well as the 'efficacy of humans and non-humans within the greater whole of a scenographic event' (7). As she explained:

> *Material Matters* drew my bodily immersion with space and materials towards a deeply resonant interactive experience…I come to realise that the cloth and I are actants along with another co-creator, the space in which we move…My approach is not bound to the human body as the initiator of the event, but that of the body of space and its reciprocal flow and interaction with human and nonhuman bodies. (12, 13, 15, 17)

Louw's scenographic performance reflects a transition from the scenographer as manipulator to that of mediator, facilitator and co-creator. For performance artist Minty Donald (2014b, 2) this is about 'giving 'voice' to the more-than-human, acknowledging its liveliness and potential agency'. As Donald (2014a, 119) contends, 'this reframing gives due consideration to the performance of all stuff: human and nonhuman; on molecular and cosmic scales; within timeframes measured in nanoseconds and eons'. Connecting with the vibrancy of the more-than-human world can aid in resisting 'culturally conditioned understandings of human/environment relations' such that scenographers might 'reimagine constructive and critical models of human/environment interdependency … thus offering speculative, inspirational alternatives … to dominant ecological narratives of disaster, human powerlessness, and blame' (119, 130). Here, Donald explicitly highlights that engaging with sustainability in scenography need not be framed around notions of reduction and limitation but can seek to connect with the living world in a way that is proactive, positive and creatively inspiring.

While theatre's engagement with the material turn is not necessarily explored in the context of sustainability, the re-interrogation of human-material relationships opens up an alternative perspective for scholars and practitioners to embrace wider ecological ideas that highlight the 'flow of imaginative exchange' (Louw 2013, 7). Ecomaterialist ontologies open up the 'worlding' of materiality in the broader context of the more-than-human world, encouraging theatre practitioners to see themselves as part of the ecological effects (both positive and negative) that they are creating, including considerations of emergent socio-political and environmental impacts in and beyond the production.

At the crux of Ecoscenography is the notion of co-designing in partnership with more-than-human collectives. As McKinney (2015a, 90) notes, it is by acknowledging 'the capability of things themselves, human and non-human alike' that we can engage with a more expansive and 'provocative frame of ideas' in scenography. If Ecoscenography enhances the designer's relationship with the more-than-human world, could this be the path that moves theatre towards 'a more ecological and more materially sustainable mode of production and consumption' (Bennett 2010, ix)? Could deconstructing the binaries between humans and non-humans (or more specifically, between scenographers and materials) lead practitioners to consider all matter as 'alive', an active part of the ecosystem, part of the web of life? And if so, how might Ecoscenography promote new sensibilities and responsibilities that alter the way the designer approaches their practice? These are just some of the questions that form the basis of reimagining scenography for an ecological paradigm.

From Phenomenological to Ecological and Evolutionary Aesthetics

If we are to bring an ecological ethos into performance design, an important deliberation is how this ethos applies to aesthetics. Philosopher Gernot Böhme (1993, 114, 125) argues that engaging with ecological ideas constitutes a new aesthetic field. While Ecoscenography uses ecological thinking to inform production processes, it is also inspired by Böhme's contention and explicitly seeks to interrogate how ecological practices can lead to new aesthetics and spatial experiences. To do so, we must critique our current notions of aesthetics.[7] While a 'successful' theatre work relies on its ability to enthral audiences through its visual-aural-spatial aesthetics, adopting an ecological perspective requires that we account for the social and environmental impacts of our ephemeral designs. We must engage with the often unseen causations of material entanglement—across bodies, substances and environments—well beyond the theatre. The shift brings up bigger questions of how we practise—of how we design the 'momentary spectacle' of the theatre experience, while also considering the lasting implications of such work.

Phenomenological Aesthetics

Phenomenological approaches to performance design are explored through the question of how one 'experiences' scenography through the multi-sensory somatic encounter (e.g. Garner 1994; McKinney 2008; McKinney and Butterworth 2009; Collins and Nisbet 2010; Lotker and Gough 2013; McKinney 2015b). Joslin McKinney and Philip Butterworth (2009, 4) emphasise how 'scenography is not only something to be seen by an audience but also something to be experienced'. As Sodja Lotker and Richard Gough (2013, 3) expand, 'we engage with scenography…by "observing" with our whole bodies'. Sensed knowledge is experienced through bodily perception (visual, audible, kinaesthetic, tactile, olfactory, somatic, gustatory, spatial and durational) and is also concerned with the reciprocal connection of the perceiver and object. It is 'the manner in which one is bodily present for something or someone or one's bodily state in an environment' (Böhme 1993, 125).

The connection between the seeing and the sensing body takes influence from the philosophical discourse of Maurice Merleau-Ponty and Mikel Dufrenne where spatial perception is not experienced as a separate entity but as one that is related to the 'perceiving body' and is integrated with the senses (Berleant 1991, 87). As Merleau-Ponty (1964, 178) explains, "I do not see it [space] according to its exterior envelope; I live in it from the inside; I am immersed in it". Merleau-Ponty's conception of spatial knowledge recognises the immediate connection between bodily awareness and visual perception (1962, 206). "My body", Merleau-Ponty highlights, "is not a collection of adjacent organs but a synergic system, all the functions of which are exercised and linked together in the general action of being in the world" (234). McKinney highlights how Merleau-Ponty's work encourages "a phenomenological account of perception from a non-hierarchical relationship between subject and object" thereby opening up our perceptions of the more-than-human world in scenography (2015a, 121). This connects with ideas of ecological thinking and forms the basis of the link between phenomenological perception and broader ecological concerns.

Arnold Berleant's (2000) concept of 'the aesthetic field' builds on Merleau-Ponty's philosophy of perception and focuses on the reciprocal action of interacting forces between the body and the environment. For Berleant, aesthetics do not end with the object, production or perceiver but are also affected by social contexts, ethnicities, customs, technological progress and circumstances. Berleant's participatory model is one of engagement (of the 'lived space') between the experiencing agent and the object of experience, the interaction of which informs the perceiving body. He highlights how aesthetics:

> provides a key to environmental understanding. It enables us to grasp the environment as a setting of dynamic forces, a field of forces that engages both perceiver and perceived in a dynamic unity. What is important are not physical traits but perceptual ones, not how things are but how they are experienced. In such a phenomenological field the environment cannot be

> objectified; it is rather a totality continuous with the participant. (Berleant 2005, 14)

Berleant's concept of 'environment' resonates with the ecological worldview and ecomaterialism which does not differentiate it as a separate entity that surrounds oneself. Rather, as Timothy Chandler and Kate Rigby assert, "it is through an aesthetic embodiment of perception "that we can rediscover ourselves as inherently ecological selves" (Chandler and Rigby 2010, 61). As Berleant explains:

> [Gabriel] Marcel urges us to say not that I have a body, but that I am my body. So we can say, in like manner, not that I live in my environment, but that *I am my environment*. As the body can be considered an extrapolation from the unity of the self, the environment can be regarded in much the same way. Thus the concept of the environment must be deepened to assimilate the lived-body, on the one hand, and broadened to embrace the social, on the other. (Berleant 1992, 96; my emphasis)

Here, Berleant's ideas of aesthetic engagement are fundamental to the rejection of traditional dichotomies of separation—between ecology and scenographic experience, performers, designers and audiences. Berleant explains how:

> understanding environment as a perceptual process is transformative. Environment is no longer an object, it is not surroundings, nor is it separate and apart from the human participant. Rather we recognize that the human is an integral constituent of environment, acting and re-acting as part of its constant flux. (Berleant 2014, 22)

This opens up ideas of how aesthetics might promote an understanding of ecological complexity that includes audiences as active participants in the creation of experience.

Expanding on Merleau-Ponty and Berleant's ideas of aesthetics, anthropologist Tim Ingold (2000, 268) acknowledges perception as a continuous process where the "knowledge of the world is gained by moving about in it, exploring it, attending to it, ever alert to the signs by which it is revealed". For Ingold, perception is about acknowledging ourselves as participants in the "currents of a world-in-formation" (2011, 129). Ingold's perspective is further discussed through my work on *This Is Not Rubbish* in Chap. 5.

Ecological and Evolutionary Aesthetics

While the dominant discourse in scenography has focused on phenomenological aesthetics such as those defined by Merleau-Ponty, Berleant and Ingold (see McKinney 2008; McKinney and Butterworth 2009; McKinney 2015a), Ecoscenography is interested in how an ecological approach may be incorporated into aesthetic experience. How might theatre aesthetics be expanded to incorporate a more explicit relationship with ecological values and ethics? These questions are examined through the concepts of ecological and evolutionary aesthetics as it relates to fields of sustainability, landscape and built environments.

Ecoscenography is concerned with taking a wider view of aesthetics, one which expands the 'environment' to not just that experienced during the performance, but that beyond the performance in both space and time. Ecoscenography demands a more comprehensive concept of aesthetics that considers a design's greater ecological integrity.

Yet, this 'integrity' involves more than just the tangible; it also extends to the creation of ecological awareness. 'Ecological aesthetics' is an emerging discourse that is concerned with 'the aesthetic appreciation of the world in its entirety' (Toadvine 2009, 85). For environmental philosopher, Ted Toadvine, ecological aesthetics is wide ranging and one which encompasses a phenomenological approach as well as a more intellectual concern with sustainability (Ibid.). While phenomenology is fundamental to 'merging biological and ecological concepts of sustainability with aesthetic appreciation', social scientist Paul H. Gobster (1999, 61) highlights how ecological sensibility relates to the 'intellectual and affective capacities [that] engage an individual to understand, appreciate and ultimately act upon the environment in a purposeful way'. This implies a cognitive stance to aesthetics that relies on ecological education as well as sensory experience. Simply put, ecological aesthetics asks us to consider how something can possibly be perceived as 'beautiful' if it taxes and degrades living ecosystems. This aesthetic awareness is one that relies on ecological literacy as well as an understanding of sensory experience.

Environmental philosopher Allen Carlson (2000) advocates for the distinction between sensory and intellectual perception, claiming that 'scientific cognitivism' is integral to obtaining a full understanding and appreciation of ecological aesthetics. At the same time, ecological aesthetics does not regard sensory and intellectual knowledge as oppositional, but rather as a correspondence between the two that can provide a useful tool for ecological decision-making. For example, Mark DeKay's (2011) *Five Stages of Sustainable Design Aesthetics*—visual, phenomenological, process, ecological and evolutionary—demonstrates how sensorial perception can coincide with a wider perspective of ecological aesthetics in a way that is not opposing in concept, but is integrated into a more holistic means of perception. Here, an ecological aesthetic is based on both an educated understanding of ecological complexity and a focus on aesthetic experience and appreciation.

Ecological aesthetics refers to an appreciation for things that are aligned with environmental health where 'perceived beauty' is 'integrated at the biospheric level' (Gobster 1999, 61). In the field of landscape research, Gobster explains how an ecological aesthetic engages with 'expanded thinking'—the way in which one experiences 'the landscape and one's place in it', where aesthetic 'pleasure is derived from knowing how the parts of the landscape relate to the whole' (1995, 8). This also relates to DeKay's notion of 'evolutionary aesthetics', where 'something can be considered beautiful if it reveals how it changes and evolves over time, especially toward greater integration, order and complexity' (2011, 119).

Both ecological and evolutionary aesthetics refer to taking a longer view of structures, landscapes, ecosystems, patterns and processes which requires an 'instructed eye' of ecological awareness and extended perception (DeKay 2011, 117). This refers to a reorientation of 'our aesthetic judgement to focus on the ecological integrity of a material, object or experience—to reconsider our appreciation for things based on a greater understanding and concern for social-ecological systems' (Beer 2016b, 489). In particular, evolutionary aesthetics' focus on appreciation of change over decades and centuries requires a heightened understanding of ecological complexities that necessitates capacities for memory, education and long-term commitment. These notions of aesthetics are not only highly intellectual and abstract, but also extremely problematic in the temporary landscape of theatre which prides itself on ephemerality and transitory engagement. Nevertheless, considering an expanded notion of aesthetics that includes phenomenological, ecological and evolutionary perspectives is fundamental to activating Ecoscenography. Long-term ecological perspectives towards a work of design and its processes—where the success of a work is measured not only by its ephemeral or 'celebratory' outcome (the performance season) but also by its ongoing potential or ecological value—have significant implications for our field.

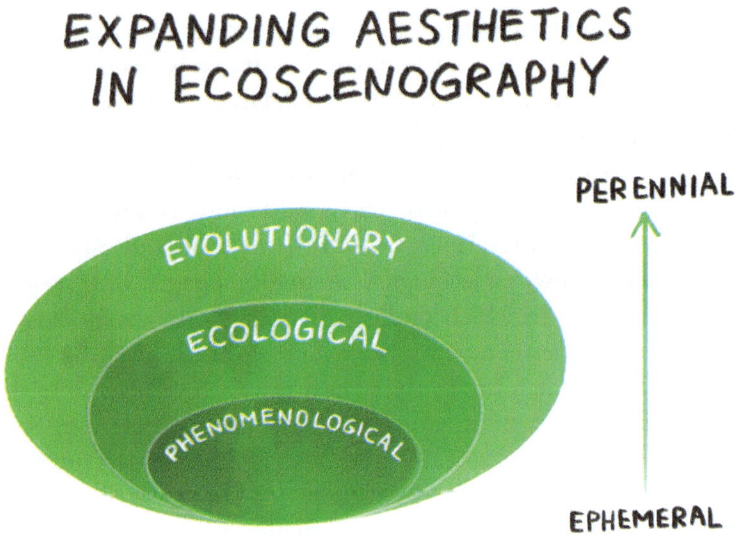

Expanding aesthetics ultimately requires that theatre makers have an appreciation of the process and evolution of a work beyond its phenomenological aesthetics, traditional structures and expectations, one that takes into account its ephemeral and perennial (long-term) concerns. In Ecoscenography, choosing one scenic material over another involves a complex understanding of its phenomenological and ecological qualities and capacities, where both viewpoints are balanced in relation to one another. This requires that ecological thinking is fundamentally positioned within the creative practice itself, equal to the spatial, budgetary, and health and safety parameters of theatre design. A key consideration of this shift will involve cogitating ways in which designers can better integrate ecological ethics and aesthetics with (the more visible) phenomenological experiences, including informing audience members more explicitly of this process. For spectators who increasingly scrutinise the integrity of everyday items, such as electricity, food, cosmetics and fashion, the need to acknowledge the socio-ecological impact of that which we consume culturally seems highly plausible.

Ecological Thinking and Creativity in Practice

Ecoscenography is an eco-creative practice which emphasises the theatre maker's imaginative capacity to place sustainability and 'its principles of resilience and strength, creative regeneration, and respect for the earth' at the core of the artistic vision (Tickell 2012). This merger allows creativity to inform ecological thinking as much as ecological thinking informs creativity—one that highlights inventive, inspiring and engaging ways of thinking and doing (Beer and Hes 2017). For London-based independent theatre company Fast Familiar (originally FanSHEN), bringing an ecological ethic into theatre production is a case of saying ecological design is my 'starting point', rather than this is limiting me as an artist (Briscoe 2013). Creative directors Dan Barnard and Rachel Briscoe explain that sustainability as a superficial 'nice-to-have' or behind-the-scenes 'add-on' inhibits the potential for creativity to inform ecological practice. Instead, 'doing sustainability' is what makes them better theatre makers because it can enable creative solutions that also help unify other production elements (Barnard and Briscoe 2012).

Similarly, for award-winning lighting designer Paule Constable, adopting an ecological approach does not inhibit creativity; it liberates it. Constable's view on sustainable lighting design is about clarifying choices, a confidence to strip the design back to the essence of the narrative or concept. As she explains:

> My favourite thing about lighting design is darkness. I really love an absence of light. I think it's what makes light more interesting. So often, good lighting is about what you don't put light on. Everything for me starts with nothing. I keep banging onto people about how taking responsibility for the use of resource[s] is good design practice because it implies a rigour. Rigour is my other complete obsession. Making it dramaturgical…There's a school of thought where lighting can be seen as decorative and it can be seen as 'stuff'. I think a lot of people are drawn to lighting because they're excited about working with technology and that sort of thing. For me, lighting isn't about the 'stuff', it's about the air that you press. (Constable in Boyd 2019)

Taking a minimalist approach is one of many examples of how to combine ecological thinking with theatre making. Constable's work is about the subtlety of design that is dramaturgical in its focus—the idea that a narrative functions most effectively in spaces without the distraction of unnecessary objects and decorations. As scenic designer R.B. Schlather highlights, engaging with ecological thinking can prompt designers to focus 'on one central design choice that can solve many problems and communicate a great deal' (Schlather in Pickard 2013). This idea of discarding the 'unnecessary' is also an important component of Fast Familiar's work which favours activating the audience's imagination with more pared-back designs, rather than going for a literal representation—which they argue is less conducive to ecological design processes (Briscoe 2013).

For Constable and Fast Familiar, taking an ecological approach means paring back their productions to the most essential components, focusing more on the dramaturgical design elements that help tell the story which they argue does not sacrifice the 'magic' and 'quality' of their work. Having a design that solely focuses on supporting a story, where every element added is carefully considered, leads them to ask: 'Do we actually need this?' on both an environmental and aesthetic level. In the case that the item in question is deemed 'necessary', it is an active choice to add something, rather than a decision made on a whim (Barnard and Briscoe 2012). Integrating ecological thinking into the design process often encourages them to question the need for particular 'things' as part of their practice. The artists' focus is both simultaneously ecological and aesthetic. As Barnard and Briscoe (2012) reiterate:

> I once heard a literary person I deeply respect articulate the idea that every word in a script which is there but doesn't need to be is depreciating the value of all the words that do need to be there. The same is true of stuff. If you have only the stuff you really need onstage then it is all valuable. Your savings can be both environmental and artistic—a combined process. (Ibid.)

For the above artists, embracing ecological aesthetics through minimalism can lead to a higher-quality result than working without it. By allowing the ecological and phenomenological aesthetics to inform each other, the artists ask 'what do we need?' and 'what can we do without?' in the making of performance. By understanding what is important, the artists eliminate that which is superfluous, giving weight to what remains, one where every design element has been highly scrutinised for its role on stage.

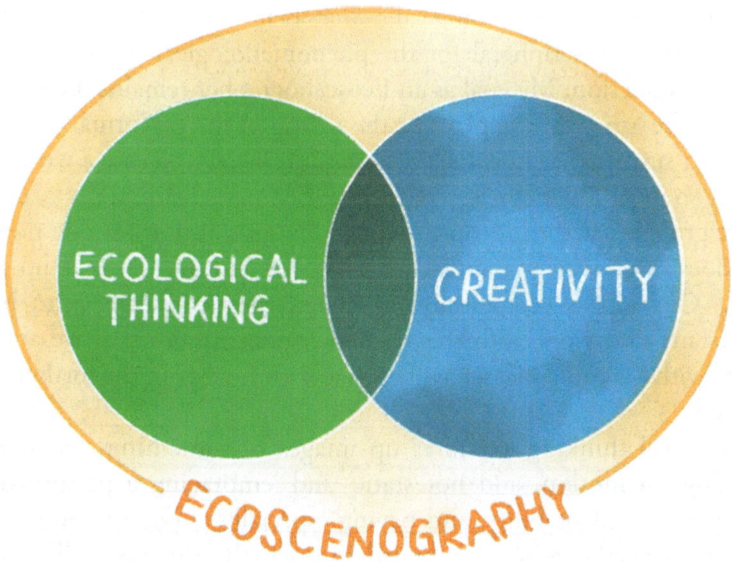

Personal Reflections on Ecological Thinking for Performance Design

In my own practice, I have found that incorporating a sustainability ethic is no different from negotiating other parameters of theatre production. As a performance designer, I am familiar with juggling a myriad of phenomenological considerations and possibilities that encompass the director's vision as well as the unique legislative, technological, budgetary, accessibility and spatial constraints that come with designing for theatre. Considering an expanded form of aesthetics has led me to gain a greater appreciation for materials and processes that I had previously overlooked. In embracing both a phenomenological and ecological approach to aesthetics, I have designed with discarded pallets, reclaimed salami netting, recycled timber, fallen tree branches and living plants. I have avoided using toxic fire retardant[8] by focusing on wearable set elements to combat stringent OH&S policies. Each of these designs has in no way hindered my ability to achieve the high-quality visual-spatial aesthetics that is demanded of my role. I have produced results that are incredibly varied in their look and feel—sometimes highly minimalist, other times more maximalist, where the ecological values behind the work range from being central to peripheral for the phenomenological outcomes of the theatre production. My goal as an Ecoscenographer remains the same: to support the 'story' of the play (or the telling of the performance), and I am often surprised at how little I sometimes need to create a world that can say so much. Nonetheless, what I have come to learn in the years that I have grappled with this topic is that a design solution for one production does in no way guarantee an answer for another. There is no silver bullet. Our job as ecologically orientated theatre makers is to better understand and work with the aesthetic opportunities that we are presented with—a topic that we will engage with further in the forthcoming chapters.

Ecological thinking 'conjures up images of something that is ever-changing, in motion, and not static' and 'embracing a perspective of openness and adaptability, and paying especially close attention to surprising discoveries that are revealed through the process' (Allen 2019,

121). Ecoscenography is profoundly 'relational'—where ecological considerations are embraced as part of the visual-spatial-aural-somatic-haptic aesthetics of theatre production. I see it as 'eco-creative dreaming', one that is systems based rather than product based, where materiality, ephemerality, sustainability and/in performance are no longer viewed as separate entities, but as relational constructs that align our 'human genius' with the 'regeneration of both our ecosystem and ourselves' (Hes and du Plessis 2014, 74). The question might now be posed as to how an ecological approach might be achieved in a theatre industry that is so rooted in its humanistic framework. How can the performing arts shift into a paradigm that is more conducive to the model of ecological thinking proposed above? And what are the ways that this might manifest in performance considerations and practices? These questions and challenges are explored in the next chapter, where the potential of ecologically based design movements and precedents provide inspiration for ecological thinking in practice.

Notes

1. Other terminologies within this discourse include 'new materialist', 'neo-materialism' and 'ecomaterialism', 'material ecocriticism' (Iovino and Oppermann 2012a); 'material feminisms' (Alaimo and Hekman 2008); 'vital materialism' (Bennett 2010); and 'agential realism' (Barad 2007).
2. The 'material turn' has a broad and interdisciplinary scope that includes feminism, political, social and economic sciences, cultural theory, anthropology, environmental philosophies, animal studies, visual arts, material culture, biosemiotics, green poetics, philosophical biology, quantum physics, deep ecology, disability, interspecies studies, gender and queer theories, geography, technology and new media. Due to its wide-ranging theories, it is not possible to detail all the movements, scholars, terminologies and theories of the field in this book.
3. Andrew Pickering refers to a 'posthumanist space' in which the human and 'more-than-human' are constitutionally 'mangled' or 'entangled' (Pickering 1995, 26). 'More-than-human' is a term from David Abram (1997) that resists reinforcing human and non-human binaries. Posthumanist scholars include Karen Barad (2003) and Donna Haraway

(2008). For more information on Posthumanism, see Francesca Ferrando (2013).

4. For example, Jordana Rosenberg (2014) argues that the ontological turn of ecomaterialisms fetishises the breaking apart of subject and object binaries.
5. Jane Bennett has adopted this concept from Bruno Latour. In *Politics of Nature*, Latour (2004, 237) defines an 'actant' as an agentic force that can be human or non-human with the capability to produce effects and influence the course of events. Bennett also refers to this as 'thing-power'.
6. Bennett (2010, 22–24) refers to assemblages as ad hoc groupings of diverse elements, with uneven topographies, not governed by a central force. Also see Gilles Deleuze and Félix Guattari's *A Thousand Plateaus* (2004).
7. By my use of the word 'aesthetics' I am implying a contemporary understanding of the term which departs from Immanuel Kant's (1724–1804) and other Enlightenment scholars' (e.g. David Hume 1711–1776; Edmund Husserl 1859–1938; Martin Heidegger 1889–1976) hierarchical preoccupation with 'taste', 'visual appearance' and 'contemplative distancing' to that which encompasses a broader scope of sensory experience and perception. Here, the epistemology of aesthetics is returned to its Greek etymological origins ('aisthēsis') which translates as 'perception by the senses' (Berleant 2014, 18) and implies a more fundamental engagement with the experience itself.
8. Flame retardants such as polybrominated diphenyl ethers or PBDEs are organobromine compounds linked to health hazards such as cancer, lung and kidney disease, reproductive disorders, birth defects and decreased nervous system development. In *Poisoned Planet: How Constant Exposure to Man-Made Chemicals Is Putting Your Life at Risk* (2014), Julian Cribb explains how manmade chemicals move rapidly in time and space—travelling on the wind, in water, soil, dust, combining with particles and influencing ostensibly natural global food chains that can have effects as far away as polar bears in the Arctic.

References

Abram, David. 1997. *The Spell of the Sensuous: Perception and Language in a More-than-human World*. New York, USA: Vintage Books.

Alaimo, Stacy. 2010a. *Bodily Natures: Science, Environment, and the Material Self*. Indiana, USA: Indiana University Press.

———. 2010b. The Naked Word: The Trans-Corporeal Ethics of the Protesting Body. *Women & Performance* 20 (1): 15–36.

———. 2011. New Materialisms, Old Humanisms, or, Following the Submersible. *Nordic Journal of Feminist and Gender Research* 19 (4): 280–284.

———. 2012. Sustainable This, Sustainable That: New Materialisms, Posthumanism, and Unknown Futures. *PMLA* 127 (3): 558–564.

Alaimo, Stacy, and Susan Hekman. 2008. *Material Feminisms*. Indiana, USA: Indiana University Press.

Allen, Tania. 2019. *Solving Critical Design Problems: Theory and Practice*. New York, USA: Routledge.

Allen, Jess, and Bronwyn Preece. 2016. Decentring the Stage: Towards an Ecocentric Ethic of Performance. *Performance Ethos: International Journal of Ethics in Theatre & Performance* 5 (102): 3–15.

Arons, Wendy. 2012. Queer Ecology/Contemporary Plays. *Theatre Journal* 4: 565–582.

Arons, Wendy, and Theresa May. 2012. *Readings in Performance and Ecology*. New York, USA: Palgrave Macmillan.

Barad, Karen. 2003. Posthumanist Performativity: Toward an Understanding of How Matter Comes to Matter. *Gender and Science: New Issues* 28 (3): 801–831.

———. 2007. *Meeting the Universe Halfway: Quantum Physics and the Entanglement of Matter and Meaning*. North Carolina, USA: Duke University Press.

Barnard, Dan, and Rachel Briscoe. 2012. Perspectives from a Rowing Boat on Starting Points and New Beginnings. *FanSHEN Theatre*. https://fanshentheatre.wordpress.com/2012/10/27/perspectives-from-a-rowing-boat-on-starting-points-and-new-beginnings/.

Bateson, Gregory. 1972. *Steps to an Ecology of Mind*. Chicago, USA: University of Chicago Press.

Beer, Tanja. 2016a. Ecomaterialism in Scenography. *Theatre and Performance Design* 2 (1–2): 161–172.

———. 2016b. Reimagining the Ruins of Scenography. *ASAP Journal (Association for the Study of the Arts of the Present)* 1 (3): 487–511.

Beer, Tanja, and Dominique Hes. 2017. Sustainability in Production: Exploring Eco-Creativity within the Parameters of Conventional Theatre. *Behind the Scenes: Journal of Theatre Production Practice* 1 (1).

Bennett, Jane. 2010. *Vibrant Matter: The Political Ecology of Things*. London, UK: Duke University Press.

Berleant, Arnold. 1991. *Art and Engagement*. Philadelphia: Temple University Press.

———. 1992. Aesthetic Perception in Environmental Design. In *Environmental Aesthetics: Theory, Research, and Application*, ed. Jack Nasar, 84–101. Cambridge: Cambridge University Press.

———. 2000. *The Aesthetic Field: A Phenomenology of Aesthetic Experience*. Christchurch: Cybereditions.

———. 2005. *Aesthetics and Environment: Theme and Variations on Art and Culture*. Aldershot: Ashgate Publishing Ltd.

———. 2014. Environmental Sensibility. *Studia Phaenomenologica* 14: 17–23.

Böhme, Gernot. 1993. Atmosphere as the Fundamental Concept of a New Aesthetics. *Thesis Eleven* 36: 113–126.

Bottoms, Steven, Aaron Franks, and Paula Kramer. 2012. On Ecology. *Performance Research* 17 (4): 1–4.

Puig de la Bellacasa, Maria. 2016. Ecological Thinking, Material Spirituality, and the Poetics of Infrastructure. In *Boundary Objects and Beyond: Working with Leigh Star*, ed. Geoffrey Bowker, Stefan Timmermans, Adele Clarke, and Ellen Balka, 47–68. Massachusetts, USA: MIT Press.

Boyd, Alice. 2019. Paule Constable: Pressing Air. *Staging Change*. https://www.stagingchange.com/post/paule-constable.

Briscoe, Rachel. 2013. Theatre in a Giant Emmental Powered by Exercise Bikes—Why Ever Not? *The Guardian*. https://www.theguardian.com/culture professionals-network/culture-professionals-blog/2013/sep/13/cheese-fanshen-sustainable-theatre.

Calma, Tom. 2010. Respect, Tolerance and Reconciliation Rather than Opposition and Denial: Indigenous Spirituality, Land, and the Future of Religion in Australia. *Pacifica: Australian Theological Studies* 23 (3): 322–336.

Capra, Fritjof. 1996. *The Web of Life*. Melbourne, Australia: Anchor Books.

Carlson, Allen. 2000. *Aesthetics and the Environment: The Appreciation of Nature, Art and Architecture*. London, UK: Routledge.

Chandler, Timothy, and Kate Rigby. 2010. Introduction: Gernot Bohme, 'The Space of Poetry'. *PAN: Philosophy Activism Nature* 7: 61–62.

Clayton, Anthony, and Nicholas Radcliffe. 1996. *Sustainability: A Systems Approach*. London, UK: Earthscan Publications Limited.

Code, Lorraine. 2006. *Ecological Thinking: The Politics of Epistemic Location*. Oxford, UK: Oxford University Press.

Cohen, Jeffrey Jerome, and Lowell Duckert. 2013. *Elemental Ecocriticism: Thinking with Earth, Air, Water, and Fire*. Minnesota, USA: University of Minnesota Press.

Collins, Jane, and Andrew Nisbet. 2010. *Theatre and Performance Design: A Reader in Scenography*. London, UK: Routledge.

Cribb, Julian. 2014. *Poisoned Planet: How Constant Exposure to Man-Made Chemicals is Putting Your Life at Risk*. Crows Nest, Australia: Allen & Unwin.

DeKay, Mark. 2011. *Integral Sustainable Design: Transformative Perspectives*. London, UK: Earthscan Publications Limited.

DeLanda, Manuel. 1997. *A Thousand Years of Nonlinear History*. New York, USA: Zone Books.

Deleuze, Gilles, and Felix Guattari. 2004. *A Thousand Plateaus: Capitalism and Schizophrenia*. Minneapolis, USA: University of Minnesota Press.

Donald, Minty. 2014a. Entided, Enwatered, Enwinded: Human/More-than-Human Agencies in Site-Specific Performance. In *Performing Objects and Theatrical Things*, ed. Marlis Schweitzer and Joanne Zerdy. Basingstoke, UK: Palgrave Macmillan.

———. 2014b. Guddling About: Experiments in Vital Materialism with Particular Regard to Water. *The Goose* 13 (1): 1–26.

Drengson, Alan. 1995. *The Practice of Technology: Exploring Technology, Ecophilosophy, and Spiritual Disciplines for Vital Links*. New York, USA: SUNY Press.

du Plessis, Chrisna, and Peter Brandon. 2015. An Ecological Worldview as Basis for a Regenerative Sustainability Paradigm for the Built Environment. *Journal of Cleaner Production* 109: 53–61.

du Plessis, Chrisna, and Raymond Cole. 2011. Motivating Change: Shifting the Paradigm. *Building Research and Information* 39 (5): 436–449.

Esbjörn-Hargens, Sean. 2009. *Integral Ecology: Uniting Multiple Perspectives on the Natural World*. Colorado, USA: Shambhala Publications.

Ferrando, Francesca. 2013. Posthumanism, Transhumanism, Antihumanism, Metahumanism, and New Materialisms: Differences and Relations. *Existenz* 8 (2): 26–32.

Frost, Samantha, and Diana Coole. 2010. *New Materialisms: Ontology, Agency, and Politics*. North Carolina, USA: Duke University Press.

Garner, Stanton. 1994. *Bodied Spaces: Phenomenology and Performance in Contemporary Drama*. New York, USA: Cornell University Press.
Gobster, Paul. 1995. Aldo Leopold's Ecological Aesthetic. *Journal of Forestry* 93 (2): 6–10.
———. 1999. An Ecological Aesthetic for Forest Landscape Management. *Landscape Journal* 18 (1): 54–64.
Goldsmith, Edward. 1992. *The Way: An Ecological Worldview*. New York, USA: Cornell University.
Haraway, Donna. 2008. *When Species Meet*. Minnesota, USA: University of Minnesota Press.
Hawken, Paul, Amory Lovins, and Hunter Lovins. 1999. *Natural Capitalism: The Next Industrial Revolution*. London, UK: Earthscan Publications Limited.
Heddon, Deirdre, and Sally Mackey. 2012. Environmentalism, Performance and Applications: Uncertainties and Emancipations. *Research in Drama Education* 17 (2): 163–192.
Heim, Wallace. 2014. The Continuing Lure of Ecological Performance. In *Landing Stages: Selections from the Ashden Directory of Environment and Performance 2000–2014*, ed. Wallace Heim and Eleanor Margolies, 5–8. London, UK: Crinkle Crankle Press.
Heise, Ursula. 2008. *Sense of Place and Sense of Planet: The Environmental Imagination of the Global*. Oxford, UK: Oxford University Press.
Hes, Dominique, and Chrisna du Plessis. 2014. *Designing for Hope: Pathways to Regenerative Sustainability*. Hoboken, USA: Taylor and Francis.
Ingold, Tim. 2000. *The Perception of the Environment: Essays on Livelihood, Dwelling & Skill*. New York: Routledge.
———. 2011. *Being Alive: Essays on Movement, Knowledge and Description*. Oxfordshire, UK: Taylor & Francis.
———. 2012. Toward an Ecology of Materials. *Annual Review of Anthropology* 41 (1): 427–442.
Iovino, Serenella, and Serpil Oppermann. 2012a. Material Ecocriticism: Materiality, Agency, and Models of Narrativity. *Ecozon European Journal of Literature Culture and Environment* 3 (1): 75–91.
———. 2012b. Theorizing Material Ecocriticism: A Diptych. *Interdisciplinary Studies in Literature and Environment* 19 (3): 448–475.
Jones, Ellen. 2014. *A Practical Guide to Greener Theatre: Introduce Sustainability Into Your Productions*. New York, USA: Focal Press.
Kegan, Robert. 1982. *The Evolving Self: Problem and Process in Human Development*. Massachusetts, USA: Harvard University Press.

Kramer, Paula. 2012. Bodies, Rivers, Rocks and Trees: Meeting Agentic Materiality in Contemporary Outdoor Dance Practices. *Performance Research: A Journal of the Performing Arts* 17 (4): 83–91.

Latour, Bruno. 2004. *Politics of Nature: How to Bring the Sciences into Democracy*. Massachusetts, USA: Harvard University Press.

Lotker, Sodja, and Richard Gough. 2013. On Scenography: Editorial. *Performance Research: A Journal of the Performing Arts* 18 (3): 3–6.

Louw, Illka. 2013. *From Designer Through Space to Spectator: Tracking an Imaginative Exchange Between the Actants of a Scenographic Event*. Cape Town, South Africa: University of Cape Town.

Mang, Pamela, and Bill Reed. 2015. The Nature of Positive. *Building Research and Information* 43 (1): 7–10.

McHarg, Ian. 1969. *Design with Nature*. New York, USA: American Museum of Natural History.

McKinney, Joslin. 2008. *The Nature of Communication Between Scenography and Its Audiences*. Leeds, UK: University of Leeds.

———. 2015a. Scenographic Materialism, Affordance and Extended Cognition in Kris Verdonck's ACTOR #1. *Theatre and Performance Design* 1 (1–2): 79–93.

———. 2015b. Vibrant Materials: The Agency of Things in the Context of Scenography. In *Performance and Phenomenology: Traditions and Transformations. Routledge Advances in Theatre & Performance Studies*, ed. Maaike Bleeker, Jon Sherman, and Eirini Nedelkopoulou, 121–140. Hoboken, USA: Taylor and Francis.

McKinney, Joslin, and Philip Butterworth. 2009. *The Cambridge Introduction to Scenography*. Cambridge, UK: Cambridge University Press.

McKinney, Joslin, and Scott Palmer. 2017. *Scenography Expanded: An Introduction to Contemporary Performance Design*. London, UK: Bloomsbury Methuen Drama.

Merchant, Carolyn. 1980. *The Death of Nature: Women, Ecology and the Scientific Revolution*. New York, USA: Harper.

Merleau-Ponty, Maurice. 1962. *Phenomenology of Perception*. New York: Humanities Press.

———. 1964. *The Primacy of Perception, and Other Essays on Phenomenological Psychology, the Philosophy of Art, History, and Politics*. Evanston: Northwestern University Press.

Morton, Timothy. 2010. *The Ecological Thought*. Massachusetts, United States: Harvard University Press.

Næss, Arne. 1989. *Ecology, Community and Lifestyle*. Cambridge, UK: Cambridge University Press.

Pascoe, Bruce. 2014. *Dark Emu, Black Seeds: Agriculture or Accident?* Broome, Australia: Magabala Books.

Pickard, Jeremy. 2013. On Eco-theater. In *Innovation in Five Acts: Strategies for Theatre and Performance*, ed. Caridad Svich, 115–125. New York, USA: New York Theatre Communication Group.

Pickering, Andrew. 1995. *The Mangle of Practice*. Chicago, USA: University of Chicago Press.

Prigogine, Ilya, and Isabelle Stengers. 1985. Order Out of Chaos. *British Journal for the Philosophy of Science* 36 (3): 352–354.

Rees, William. 1995. Achieving Sustainability: Reform or Transformation? *Journal of Planning Literature* 9 (4): 343–361.

Ribac, François. 2018. Narratives of the Anthropocene: How Can the (Performing) Arts Contribute Towards the Socio-Ecological Transition? *Scene* 6 (1): 51–62.

Riskin, Jessica. 2015. *The Restless Clock: A History of the Centuries-long Argument Over What Makes Living Things Tick*. Chicago, USA: University of Chicago Press.

Rose, Deborah Bird. 1996. *Nourishing Terrain: Australian Aboriginal Views of Landscape and Wilderness*. Canberra, Australia: Australian Heritage Commission.

Rosenberg, Jordana. 2014. The Molecularization of Sexuality: On Some Primitivisms of the Present. *Theory & Event* 17 (2).

Schumacher, Ernst Friedric. 1973. *Small is Beautiful*. New York, USA: HarperCollins.

St. Pierre, Louise. 2019. Design and Nature: A History. In *Design and Nature: A Partnership*, ed. Kate Fletcher, Louise St. Pierre, and Mathilda Tham. London, UK: Routledge.

Tickell, Alison. 2012. Sustainability Should be at the Heart of our Collective Artistic Vision. *The Guardian UK*. https://www.theguardian.com/culture-professionals-network/culture-professionals-blog/2012/oct/25/sustainability-arts-council-julies-bicycle.

Toadvine, Ted. 2009. Ecological Aesthetics. In *Handbook of Phenomenological Aesthetics*, ed. Lester Embree and Hans Reiner Sepp, 85–91. Dordrecht, Netherlands: Springer.

Woynarski, Lisa. 2015. A House of Weather and a Polar Bear Costume: Ecological Anthropomorphism in the Work of Fevered Sleep. *Performance Research* 20 (2): 24–32. https://doi.org/10.1080/13528165.2015.1026722.

———. 2020. *Ecodramaturgies: Theatre, Performance and Climate Change*. Cham, Switzerland: Palgrave Macmillan.

Woynarski, Lisa, Adelina Ong, Tanja Beer, Stephanie Beaupark, Jonah Winn-Lenetsky, Rulan Tangen, and Michelle Nicholson-Sanz. 2020. Dossier: Climate Change and the Decolonized Future of Theatre. *Theatre Research International* 45 (2): 179–208.

Yunkaporta, Tyson. 2019. *Sand Talk: How Indigenous Thinking Can Save the World*. Melbourne, Australia: Text Publishing.

3

Ecological Practice

T. Beer, *Ecoscenography*, https://doi.org/10.1007/978-981-16-7178-4_3

In the last decade there has been a subtle but noticeable shift in global conversations of sustainability, one which has seen negative (mechanistic) assumptions of sustainable practice reimagined through a lens of creativity, abundance and hope. This new direction not only advocates for the environmentally sensitive management of waste and resources but also makes space for novel possibilities and ways of working that forefront socio-ecological values. The performing arts sector—one of many others—finds itself at an exciting moment of transition, a time when sustainable approaches modelled on the ecological worldview are engendering new values, aesthetics and possibilities.

Originating more than 60 years ago, the global sustainability movement grew out of the detrimental impacts of industrial and economic growth in the twentieth century, as famously publicised in Rachel Carson's seminal text *Silent Spring* (1962). By 1987, The Brundtland Report had set out a clear directive for sustainable development, defining it as one 'that meets the needs of the present generation without compromising the ability of future generations to meet their own needs' (Brundtland Commission 1987). This notion of sustainability—balanced across three pillars of social equity, economic growth and environmental sustainability—arguably remains the most well-known definition today.

While the Brundtland Commission was instrumental in setting a global environmental agenda, it failed to shift conversations of sustainability beyond a mechanistic framework, being guided by a narrow, anthropocentric definition of what sustainability is. As discussed in Chap. 2, the rhetoric of meeting the 'needs of humans above other species' as well as the conjecture that living systems 'can be managed' reinforces people's position above nature and fails to 'take human fallibility into account' (St. Pierre 2019, 98). This 'end-game of a "doing less harm" approach' (Robinson and Cole 2015, 134) has also been contested by many sustainability experts who have argued that ecologically responsible practice needs to start by 'doing the right things', rather than 'the wrong things better' (McDonough and Braungart 2013, 17). More broadly, the idea of 'sustaining' or 'sustainment' conjures up a notion of a steady state of equilibrium which does not match, either conceptually or technically, with the way ecological systems work. Ecological systems are in a constant state of flux, surviving and thriving 'by changing and adapting, seeking dynamic equilibrium with their evolving environments' (Mang and Haggard 2016, xxv). Nevertheless, the term 'sustainability' has come to represent environmentally conscious practices and continues to be used (albeit with varying meanings) as a unifier of the field.

As highlighted in the Introduction, the legacy of the mechanistic worldview continues to dominate discourses of sustainable practice in the performing arts, emphasising mitigation and measurement without taking into consideration wider socio-ecological contexts, benefits and opportunities. This chapter explores models of ecological practice across the built environments and related industries, from eco-efficiency (rooted in mechanistic thinking) to the more holistic approaches framed by the ecological worldview that underpin Ecoscenography. While there are challenges to implementing these approaches in the performing arts, I contend—as I have above—that these movements offer originative avenues for theatre productions designed for an ecological paradigm.

The Limits of Eco-efficiency

Coined by the World Business Council for Sustainable Development (WBCSD) at the 1992 Earth Summit in Rio, the term 'eco-efficiency' is still the most widely used sustainability approach across industries. Originating from scientific, technological and economic influences of the Industrial Revolution, eco-efficient practice has been defined as a 'carrying capacity' approach to minimising waste, pollution and natural resource depletion while still satisfying human needs (DeSimone and Popoff 1997). It is largely focused on economic (not ecological) value-adding and does not seek to achieve positive environmental or social outcomes. The appeal of eco-efficiency is that it works within a business-as-usual context and is therefore more straightforward to implement in conventional circumstances. However, it has also led many to see sustainability as a burden foisted on productions through material limitations or policy restrictions, rather than one that fosters new ways of doing things.

In recent years, the performing arts has shifted to include eco-efficient strategies for theatre buildings and stage productions. Organisations such as Julie's Bicycle (UK), Creative Carbon Scotland (UK), Live Performance Australia and Broadway Green Alliance (US) are just some of many organisations worldwide that have been instrumental in developing measurement tools and offering workshops and seminars on 'how to do sustainability' from an eco-efficient perspective. Performing arts organisations such as Arcola Theatre (UK) and Sydney Theatre Company (Australia) have demonstrated how eco-efficiency has resulted in significant improvement in theatre venues—including how careless energy expenditure and excess waste can be mitigated through the use of LED lighting, locally sourced materials and reuse.

Aiming for eco-efficiency in energy usage and waste management in all sectors is vital. It is the crucial first step by which the performing arts can embrace sustainability. Yet, many argue that eco-efficiency is just that—a starting point from which a more holistic approach to sustainability can be pursued. Given the harsh realities of environmental degradation and climate change, leading sustainability experts contend that we must do more than simply minimising negative impacts. For example, Janis Birkeland (2012, 163–164) argues that eco-efficiency ignores the fact that development has already exceeded the earth's 'ecological carrying capacity' and, thus, merely slows down degradation.

Eco-efficiency has traditionally been focused on a quantitative approach to measurement and procurement (mechanistic worldview) without activating social, cultural and environmental potential (ecological worldview). A holistic socio-ecological approach to sustainability cannot be achieved if we continue to relate to natural systems with overly simplistic terms of 'impact reduction' or 'limiting the damage' (Eisenberg and Reed 2003, 6). Similarly, the emphasis on quantification produces an emphasis on predicting and assessing negative impacts, rather than aspiring to strategies that can create positive impacts long term (Birkeland 2012, 80). Mechanistic thinking, focused on technological, industrial and economic systems, undermines the development of more innovative possibilities in sustainable practices (Eisenberg and Reed 2003, 1). It ignores the cultural, social and economic complexity that underpins our environmental issues and undervalues our creative and contributive capacity as humans.

Eco-efficiency is also limited in its largely colonial, homogeneous, white, patriarchal perspectives that forefront a notion of sustainability as '"green and pleasant", middle class, white, singular and reductive' (Woynarski 2020, 36). To recognise these inequalities and prejudices, we need 'an intersectional understanding' of sustainability (see Chap. 2) that exposes 'the complexity and interlinked issues of oppressions and injustices' (34), including 'how the ideologies of colonialism and imperialism are still felt through oppressive social structures' today (52). Truly sustainable design should work towards regenerating more-than-human places by intersecting with and acknowledging a diversity of communities and perspectives, a step well beyond focusing only on resource mitigation.

There is also evidence to suggest that eco-efficiency's emphasis on mitigation has induced apathy, rather than active engagement on sustainability issues (Nordhaus and Shellenberger 2009). As McDonough and Braungart (2002) surmise, defining our lives in the negative, focused on 'what not to do' or forcing old paradigms to be 'less bad', does not motivate change. Eco-efficiency's emphasis on 'limiting' or 'restricting' practice also constitutes a lack of vision and inspiration that is not conducive to creative industries that pride themselves on artistic integrity and innovation. Regularly regarded as 'tedious' or 'boring', engaging with sustainability in the performing arts is commonly regarded as a chore. In this context, environmentally responsible strategies are often incorporated after the design concept has been conceived; they have not been integral to the scenographer's ideas, processes and aesthetics.

Despite often being viewed as a burden, the eco-efficiency model has been popular in the performing arts largely because it does not challenge the current status quo of operations (Brunner and Mehler 2013, 26). While I recognise eco-efficiency as a vital step, I suggest that this model—on its own—limits opportunities for creative potential in Ecoscenography. Theatre needs a socio-ecological approach that it can aspire to, one that uses eco-efficient tools to help understand and mitigate our impacts, but aims higher, to pursue work that not only intersects with broader notions of inequity but is also 'abundant, prosperous and intelligent from the start' (McDonough and Braungart 2002, 75).

Towards an Ecological Approach

Ecoscenography is founded on a number of ecological design movements that evolved from the grassroots cultures of the 1960s and 1970s and established new modes of thought within the field of built environments. These include biophilic design (Kellert et al. 2008), circular design (Webster 2017), eco-effective design (McDonough and Braungart 2002), bio-inspired design (Benyus 2002), ecological design (Van der Ryn and Cowan 1996) and regenerative development (Reed 2007). The sequence of the above design movements can be seen as an evolutionary path in shifting from a human-centric (mechanistic) to a whole systems (ecological) approach. I summarise their key values and potential for Ecoscenography below, including examples of how a variety of designers, artists and theatre makers have used these approaches in their own work. Despite the fact that ideas of ecological design are founded on more holistic concepts of the ecological worldview (Chap. 2), I acknowledge that there is an urgent need for a greater expansion of scholarship and practice that crosses ethnic, racial and geographic divides—particularly those that bring more diversity to these potentialities, especially within marginalised communities.

Biophilic Design

The concept of biophilia—literally translated as 'love of life'—was popularised by biologist Edward O. Wilson (1984) to highlight the inherent biological affiliation between humans and the living systems in which we are part. It is the idea that humans have an innate propensity to seek connections with nature and other forms of life. The biophilia hypothesis is based on humanity's enduring co-evolvement with nature and the necessity of more-than-human connectedness to foster physical, emotional and psychological wellbeing (Kellert et al. 2008). Many of us will be familiar with how nature-based activities, such as a walk in a rainforest, a swim in the ocean, listening to birdlife, tending to our plants or access to natural light, make us feel better. This affiliation can have a remarkable effect on our mental health, including lowering our stress levels and bringing us into the present moment. It is the reason why many of us choose to have plants in our homes, seek out views of natural landscapes or prefer to wear natural (rather than synthetic) fibres.

Biophilic design is the conscious creation of environments that strengthen the integration and connection of living systems through designer-based interventions (Kellert et al. 2008). It encourages designers to think about how reciprocal benefits between the built environment and the natural environment can be sustained and achieved. The purpose of biophilic design can be threefold: (1) to 'enhance people's love of nature', (2) to 'inform people about nature' or (3) to 'offer nature as a service to improve human life' (St. Pierre 2019, 100). Designers may use natural shapes, forms, patterns, processes and light that foster a sense of place or connection to the outdoors, such as organic shapes to bring nature into an interior environment or spaces that open up place-based relationships through access to local materials and finishes. For example, biophilic design inspired the Willow Theatre, created for World Stage Design 2013 at the Royal Welsh College of Music & Drama in Cardiff.[1] Architect Tim Lai and stage designer Brad Steinmetz produced a theatre constructed entirely from locally sourced, reusable and recyclable materials, including a hired scaffolding frame covered with horticultural fleece. The designers were inspired by the fleece material itself, which was readily available and easily recyclable, with a light, airy, translucent quality. By hanging the translucent fabric strips around the scaffolding, Lai and Steinmetz were able to create a sense of a 'living theatre' in which the moving fabric responding to the natural breeze resembled the swaying branches of a willow tree.

Biophilic Design in Scenography

The notion of biophilic design in scenography is not new. Theatre makers all over the world and across time have used the natural landscape as a feature or site-specific backdrop to integrate performances with the more-than-human world. Theatre venues have created lavish sets that mimic natural elements and/or scenery to depict 'a feeling of the outdoors' inside. In most cases, biophilic design in the theatre is more concerned with 'serving the play or production', rather than its connection to our health and wellbeing. Its feature is one that is largely contextual, not based on a wider concern for facilitating more-than-human connection beyond the needs of the play or the performance itself. Nevertheless, biophilic design can create a visceral response that promotes a sustainable ethic, and this makes it an important tool for the Ecoscenographer.

Vietnamese performance maker Tuấn Lê chose bamboo for his circus show *À Ố Làng Phố*[2] (loosely translated as 'Village to City') because of its gentle beauty and cultural significance to Vietnam (Asia TOPA 2020). Performed at Asia TOPA (Arts Centre, Melbourne, 2020), the material evoked a new aesthetic for circus performance that was a world away from traditional apparatus, often made from synthetic fabrics, steel and timber. Instead, the bamboo set design became the 'star of the show', with its natural strength and flexibility that helped tell the story of urbanisation in Vietnam. Through the simple use of bamboo, Lê's set design immediately evoked a sense of life in South-East Asia, from its village of rattan-woven boats and woven baskets to urban construction sites. However, the warmth and authenticity of the material also provoked an instinctual connection to the natural world that extended beyond its Vietnamese setting. Whether Lê intended to or not, bamboo in *À Ố Làng Phố* was depicted as a sustainable and durable resource of great beauty.

In contrast, biophilic design was a much more deliberate choice for *Glimpsing Air Pockets*,[3] a collaborative dance theatre production with, for and by children led by choreographer Christina Liddell and Ecoscenographer Mona Kastell, with composer Sinclair and projection artists McHenery and Momenpour (Fig. 3.1). The project specifically aimed to enhance children's wellbeing through nature connection (Kastell and Myers 2019, 29). Working with children from the Edinburgh Royal Sick Kids Hospital, the creative team conducted creative workshops with participants, weaving their ideas into the multisensory scenic and choreographic elements of the work. The scenography enhanced the patterns of biophilia by replicating nature through foraged and reclaimed materials, such as willow branches, tree stumps, real plants, wool, old furniture and donated pieces of carpet that were stitched into a grass patchwork. This assemblage of things and their distinct tactile quality is highlighted in Kastell's description of the design:

Fig. 3.1 Christina Liddell performing in Mona Kastell's set design for *Glimpsing Air Pockets* (Dance Base, Edinburgh, 2019). (Photo: Mona Kastell)

> We wanted to bring [the audience] together to treat the set as an immersive participatory sensory scenography to be interacted with: from visual curiosity—willow trees to stroke, felt bees to seek, rosemary mist to smell and seeds maracas to play with along their journey within the garden…and press the mushrooms to activate sounds and projections. (36)

Glimpsing Air Pockets demonstrated how biophilic design in theatre can accentuate nature connection, using both 'real' and fabricated nature, and how this can be achieved in an environmentally sensitive way, prioritising found objects and recycled materials that co-emerge from place. The project showcased how theatre can be a unique platform to bring art and nature together, to create unique biophilic environments that transport children into 'the outdoors' while producing a safe and controlled space for these experiences to occur. Here, Ecoscenography becomes one of creating biophilic encounters for those that have the least access to it, one which resonates positively long after the event.

While biophilic design can promote a greater awareness and concern of nature, its focus on phenomenological aesthetics can lead designers to see 'nature as a service' rather than one of positive reciprocity. Thus, it is important that biophilia is seen through an expanded aesthetic lens that includes the ecological and phenomenological considerations outlined in Chap. 2. In my personal practice, I am acutely aware of the potential of biophilic design as an Ecoscenographer who is increasingly focused on creating environments that not only bring audiences closer to the natural world, but also allow them to deepen their understanding of biodiversity and sustainability. For me, the use of living plants, natural materials and fibres is a conscious choice in promoting a sustainable ethic, particularly within the context of expanded scenography. I see this as a subversive biophilic design tactic, one that entices audiences through the aesthetic experience to immerse participants in an ecological act. At its best, biophilic design accentuates our relational connection to living systems through phenomenological features and makes explicit the importance of enhancing and maintaining our relationship with nature (Hes and du Plessis 2014). This is often the first step towards bringing an ecological focus to our work and regenerating the social-ecological systems of which we are part.

Circular Design and Eco-effectiveness

Circular design has been heralded as 'the next big thing in design', where the focus is on creating products and services that are waste and pollution free. Based on ideas of the circular economy, circular design disrupts linear processes of 'make, use, dispose' to recover products and services at the end of their human usage and bring them back into production (Webster 2017). This 'cradle-to-cradle' approach is not only dedicated to reducing resource extraction and consumption, but also finding creative ways to reuse, repair, re-service and restore products for economic and ecological benefit (Charter 2018). Emerging from society's rampant 'throw away culture', circular design not only advocates for designing ways in which resources can be continuously renewed but also creates new business prospects and income streams.

Promoting an extended life span through repair and reuse is a key approach of the circular economy and circular design, and in this re-embraces a mentality, a skill-set and an industry that wilted during the second half of the twentieth century. For example, the recent birth of creative mending by 'repair artists' who produce sewing styles that restore old clothing in quirky and unique ways offers an alternative to the fast-fashion movement and harks back to the specialist clothes menders whose business model was destroyed by the linear, consumptive ethos of fast fashion (e.g. Lewis-Fitzgerald 2020). Closer to the theatre world, Sandra Goldmark (2020) co-founded 'Fixup', a social enterprise and tinkering shop that uses theatre skills to promote sustainable, circular modes of use. Goldmark—a theatre practitioner and educator—hires freelance theatre technicians and artisans to repair lamps, mobile phones, jewellery, clothing, furniture, glasses, accessories, backpacks and toys (amongst other things), creating training opportunities and jobs for communities and individuals.

While circular design demonstrates great potential, the labour in separating and recovering waste is costly and time-consuming if products are not designed with circularity in mind. Some 80–90% of a product's environmental impact is determined at concept stage (Graedel et al. 1995). Thus, factoring in the reusability of a product early in the design process is crucial. On strategy, 'design for disassembly' focuses on designing for material recovery, value retention and material reuse through easy deconstruction. The approach informs decisions and material choices, changing how materials are combined and layered to ensure flexibility and reversibility. Innovative office furniture company Ahrend provides an example, creating products that are designed for modularity, disassembly and life extension, allowing easy repair, and alterations and add-ons. The company offers customers 'furniture-as-a-service', rather than 'furniture-as-a-product', where clients pay a monthly fee and can return furniture items for any repairs, adjustments or replacements, leading to a reduction in waste and carbon emissions.[4] This framework is one that ultimately creates less products, but more jobs. In another strategy for furniture, designer William Warren created 'Shelves for Life'—a set of bookshelves that literally lives and dies with you, being refashioned into a coffin for your end of life stage (Cooper 2020). Somewhat macabre, but highly practical, nonetheless.

The ultimate aim of the circular economy is to move beyond the recycling economy's focus on prolonged use (e.g. storing something that ultimately ends in landfill) to advocate for constant renewal, where all products are designed for circulation. This is the focus of McDonough and Braungart's notion of 'eco-effectiveness', which is a departure from the notion that the best we can hope for is to be 'neutral' or 'less bad' in relation to the living world (Pedersen Zari and Jenkin 2010, 3–4). Coined by McDonough and Braungart in *Cradle to Cradle* (2002) and later revised in *The Upcycle* (2013), eco-effectiveness is a popular notion of the circular economy that reframes the goals and methods of sustainable design to provide beneficial environmental and social outcomes through a circular framework. Eco-effectiveness seeks designs without *any* (rather than fewer) negative environmental or social consequences. Proponents argue that this prompts designers to rethink the possibilities of sustainable practice, eliminating the common assumption that sustainability is restricted to the three 'R's' (reduce, reuse, recycle) and suggesting ways in which waste can be designed entirely out of the process. Using the example of a worm farm, the authors highlight how waste can equal food when considered through biological nutrient cycles.

'Upcycling' describes the multitude of ways in which discarded objects can be repurposed into valued artefacts (McDonough and Braungart 2013). Unlike recycling, upcycling is about 'not merely conserving the resources that went into the production of particular materials, but adding to the value embodied in them by the application of knowledge in the course of their recirculation' (Murray cited in Earley 2011, 3). The method involves upvaluing rather than just reusing, thereby enabling waste materials to increase their status as resources. McDonough and Braungart also address the danger of embodied toxins or toxicity in design processes, drawing attention to toxicities already inherent in many everyday objects, such as flame retardant in furniture, or cancerous and hormonal disruptive toxins in plastic packaging. Eco-effectiveness' focus on dangerous toxins offers a valuable consideration in theatre practice because toxicity has been largely absent from most sustainability conversations in theatre production.

There are a countless social enterprises, not-for-profit organisations and designers that reconsider single-use products by transforming them into useful artefacts and services that extend their use, often with eco-effective social and ecological benefits in mind. Street Sleeper[5] is a social enterprise in South Africa that upcycles PVC advertising billboards, destined for landfill, into survival sleeping bags for the homeless that provide protection against the elements at night and double as backpacks during the day. Ananas Anam[6] turns pineapple waste into a vegan alternative to leather, while Chip[s] Board[7] uses waste from potato peelings to create a sustainable wood substitute alternative to MDF and chipboard. Unlike MDF, Chip[s] Board is easily biodegradable post-use and does not contain any toxic resins and chemicals (such as formaldehyde). These examples are just some of the many creative outcomes of industries around the world using the circular economy to inspire new processes across fashion, product design and architecture.

Circular Design and Eco-effectiveness in Scenography

Circular design approaches give value to the whole design and construction process, often engaging the consumer in this experience. However, its full potential is yet to be embraced in the performing arts. In the theatre, truss systems, velcro straps and standardised rostra are examples of products that consider multiple uses. But what of larger bespoke designs that are not so transferrable? EDEOS is an eco-design tool developed by the Lyon Opera that implements a circular framework in its design and construction processes. As a holistic sustainability calculator, the tool aims to assess the production of stage sets through its impact on climate, human health, ecosystems and non-renewable resources. What makes EDEOS unique is its focus on decision-making throughout the entire life-cycle of a set design, including the concept and post-production phase. As Thierry Leonardi (2020) highlights:

> As well as calculating the footprint of set designs, EDEOS also measures key indicators of eco-design, such as the percentage of reused or recycled elements used in scenic manufacturing. Eco-design indicator values are calculated before the set is constructed, as well as considering what happens to the sets afterwards based on manufacturing scenarios. This includes identifying the impact of construction methods on the quality of materials.

Unlike more quantitative or mechanistic carbon calculator formats, EDEOS' aim is not to 'have a very precise value of a set design footprint but to improve eco-design practices by making better informed decisions' (Leonardi 2020). While the project is currently designed for the Lyon Opera with its own specific conditions and supplies, it is hoped that EDEOS could become a shared industry tool that might also extend beyond the theatre to other related industries in the future.

Circular design's focus on renewable energy has also been a feature of many Ecoscenographic productions. From kinetic to solar energy, theatre makers have been investigating how lighting and sound can be powered through off-the-grid performances. Choreographer Prue Lang's work on *Un reseau Translucide* (Festival Faits d'hiver, Paris, 2011) investigated 'human activity as a renewable source of energy' to rethink 'habitual modes of theatre production' through 'an autonomous performance' that was entirely powered by the performers (Lang 2020). Lang collaborated with 'engineer-inventors' to create the renewable energy design and corresponding choreography with the goal of investigating how movement and electricity could facilitate new forms of artistic expression.[8] Similarly, Ecoscenographer Ian Garrett worked with Asethetec Studio and choreographer William Yong in *vox:lumen*[9] (Zata Omm, Toronto, 2015) to explore how lighting and sound for dance performances could be powered through movements by the dancers, the audience and renewable sources (Garrett 2016) (Fig. 3.2). These works demonstrate the diversity of circular design as one that is transferrable across scenographic disciplines.

The circular economy's focus on 'all the R's'—rethink, recycle, refuse, recover, regift, repair, reuse, reduce, reimagine—creates opportunities for eco-creativity. It allows designers to imagine a future as a 'world of abundance' rather than one of 'sustaining limits', where looking towards a

Fig. 3.2 Dancers in *vox:lumen* (Toronto, Canada, 2015). (L to R) Daniel McArthur, Brendan Wyatt, Irvin Chow, William Yong, Michael Caldwell. (Design by Ian Garrett; Photo: David Hou)

positive or thriving future becomes the inspiration. The goal of circular design and eco-effectiveness is not to minimise the cradle-to-grave flow of materials in scenography (as with eco-efficiency), but to generate cyclical, cradle-to-cradle 'metabolisms' that enable materials to maintain or increase their status as resources (upcycling) after the performance season. This implementation of cyclic thinking is fundamental to bringing an ecological ethos into performance design—something that we will explore further in the next chapter.

Circular thinking and design have become a core piece of my own practice as an Ecoscenographer working across conventional and expanded contexts. Every material choice that I make is considered within a cyclic framework that takes its 'after-performance-life' into consideration. My process often begins by investigating a theatre company's pre-existing stock (e.g. rostra, furniture pieces and staircases) to inspire possibilities that will inform the design. When considering a possible purchase, I always attempt to judge a material or product's storage capacity and potential for reuse. I have sought to eliminate mixed-material fabrication and designed for easy disassembly, storage and future use. While I often stumble on my path towards circularity, each project is an opportunity to learn new tools and strategies and extend networks that further my practice as an Ecoscenographer.

Bio-inspired Design and Bio-integrated Design

There is no such thing as waste in biological systems. Emulating this idea and the broader processes found in nature is increasingly used as a strategy for creating solutions to human challenges. Bio-inspired design—also called biomimetics or biomimicry—imitates nature's processes for novel, innovative applications. It is a transdisciplinary approach that uses observations of nature to inspire time-tested forms, patterns, processes and strategies that can help inform human solutions based on the wisdom and adaptation of 3.8 billion years of evolution that has produced living systems. In *Biomimicry: Innovation Inspired by Nature* (2002), Janine Benyus defines bio-inspired design as a practice that examines nature's greatest ideas and then imitates or mimics these ecological processes to solve human issues. A key focus is to learn from nature, rather than determine what we can extract from it, and to stimulate design thinking based on the question of how life creates conditions conducive to its perpetuation (Pedersen Zari 2016; Benyus 2002).

Bio-inspired design uses human-made technologies to develop more effective ways to use resources. For example, using the patterns of a leaf to invent a better solar cell or a termite mound to inform heating and cooling processes in the structural design of a building or material. This entails looking to the living world through emulating 'an organism, an organism's behaviour or an entire ecosystem, in terms of forms, materials, construction methods, processes or functions' (Pedersen Zari 2015, 58). Famous examples include a Japanese bullet train that took inspiration from the shape of a bird's beak to help decrease sound pollution and the creation of Velcro which emerged out of noticing how some plants have hooks to help attach seeds to an animal's fur (and thereby achieve free dispersal).

There are numerous examples of practitioners exploring new insights and opportunities of bio-inspired design, from architecture, product design and fashion. Some of these projects extend beyond biomimicry, into bio-integrated design where 'biology is both the subject and the medium' (Myers 2012, 195). Bio-integrated design is also the topic of *Lo-TEK: Design by Radical Indigenism* (Watson 2019), which highlights the wisdom of Indigenous design by exploring thousands-of-year-old nature-based technologies that hold valuable insights for contemporary practice. Examples include a living root bridge where the Khasi people of Meghalaya's jungles (Mawlynnong Village, India) have trained rubber trees to grow organic bridges over rivers for centuries and a portable Ma'dan house structure built from local reeds in the Southern Wetlands of Iraq that can be built and dismantled in a day.

In bio-integrated design, the amalgamation of living organisms is a vital part of the design, where biological systems integrate with technological systems to create new hybrids (Myers 2012). A notable example is eco-artist Oliver Kellhammer's work, *Plastivore* (Science Gallery, Melbourne, 2019), which explored the relationship between synthetic and organic processes using 10,000 live mealworms, who ate their way through one of the world's biggest waste issues: Styrofoam.[10] Meal worms can safely bio-degrade and depolymerise the plastic into a digestible and nourishing food source within a few hours, a process that otherwise takes hundreds of years (Northover 2019). As part of an exhibition of the work, Kellhammer invited the public to contribute their own Styrofoam waste as part of the 'meal worm performance'.

Drawing inspiration from micro-algae, which thrive on daylight and air, London-based architecture firm ecoLogicStudio created a living curtain that could turn carbon dioxide into oxygen through the process of photosynthesis to combat pollution. The curtain—made from bioplastic panels that contain algae attached to a public building—was able to capture and convert about one kilogram of carbon dioxide per day—the carbon sequestering equivalent to 20 trees (Chow 2018). Algae has also been used as a way of counteracting fast fashion, with Roya Aghighi[11] creating 'Biogarmentry'—clothing made with living, photosynthetic cells in collaboration with the University of British Colombia and Emily Carr University (Block 2019b). Single-cell green algae is spun together with nano polymers to create a living and photosynthetic material. Biogarmentry is literally alive, activated by sunlight, and requiring that the wearer simply spray the item of clothing with water once a week. Through the process of photosynthesis, the clothes also contribute to the immediate environment of the wearers by helping to regulate carbon emissions and are compostable after use. Costume designers may also be inspired by the work of Elissa Brunato,[12] who has created the Bio Iridescent Sequin, a sparkling bead produced with natural cellulose rather than plastic sequins (Block 2019a). Using bioplastic based on cellulose extracted from trees, the bio-sequins can be sewn onto fabric like a customary sequin or bead. In the lighting sector, Massachusetts Institute of Technology (MIT) plant nanobionics programme has explored the potential of transforming plants into lamps by inserting a fire-fly enzyme into their leaves, opening up new innovations in off-the-grid low-level indoor lighting as well as the prospect of trees acting as self-powered streetlights (Morris 2018). This work has also been extended by some designers to investigate ways in which bio-luminescent bacteria found in jellyfish and mushrooms could create 'glow-in-the-dark vegetation' (Pallister 2014).

While these experimentations are stimulating, they also provoke an ethical question of what it means to co-create with nature. Are we really co-creating or manipulating? How do these human manipulations affect habitats for species, as well as disturbing the plant's function? Does the main objective become one of using smaller living beings in order to supply our needs in the future? As Louise St. Pierre highlights in *Design and Nature: A Partnership* (2019):

> Bio design mixes the astonishment at the workings of nature with a desire to master her…the notion of mastery aligned with creativity is an alluring force in design with nature, eclipsing the energy and self-organising capacity of nature herself. (99)

It is clear that bio-integrated design requires greater investigation and interrogation if we are to use it as an ecological approach to human production. As we move towards a more holistic understanding and integration of environmental and social justice, a key priority will be to remain sceptical of bio-practices that 'prioritize human needs, mastery, and economic growth' (101) over ecological benefit. At its best, bio-inspired design can foster a way of working like nature, generating waste-free systems and resource sharing processes that contribute to the system.

Bio-inspired and Bio-integrated Design in Scenography

The amalgamation of art and science is a crucial part of both bio-inspired design and bio-integrated design (or what I might term 'bio-opportunistic') design in scenography. Bio-opportunistic designers commonly collaborate with biologists and scientists to create their works, and this was also the case for scenographer and installation artist Pascal Leboucq, who joined forces with Erik Klarenbeek (from biodesign studio Krown Design) to create the *Growing Pavilion*, a temporary event space from mushroom mycelium, reeds and agricultural waste for the 2019 Dutch Design Week.[13] The mycelium panels that clad the timber structure of the pavilion were light, portable and well insulated for temperature and sound—ideal for performance (Pownall 2019). As well as hosting a suite of theatrical events, mushrooms from the set were harvested every day in front of an audience before being cooked and served as a meal by a local foodtruck. Inside the space, an exhibition of other bio-based artworks was on display, featuring *Manureality*[14] (furniture made from horse manure by Martijn Straatman) and *Living Skin*[15] (clothes made from mycelium, kombucha and algae by Aniela Hoitink). The total artwork of the pavilion pushed boundaries and expectations of design, programming and materiality.

Like *Living Skin*, bio-costume was the focus of Ecoscenographer Ingvill Fossheim's Masters project at Aalto University Finland (2019). Working with selected species of fungi, algae, berries and microbes, Fossheim explored bio-based materials as viable alternatives in contemporary costume design and performance making. The project included garments made from microbial cellulose, pigments and dyes made from healing berries and algae, and sustainably sourced reindeer skins and hair (Fig. 3.3). The integration of organic material into costume design has also been a focus of Luanna Jimenes,[16] whose work used her clothed body

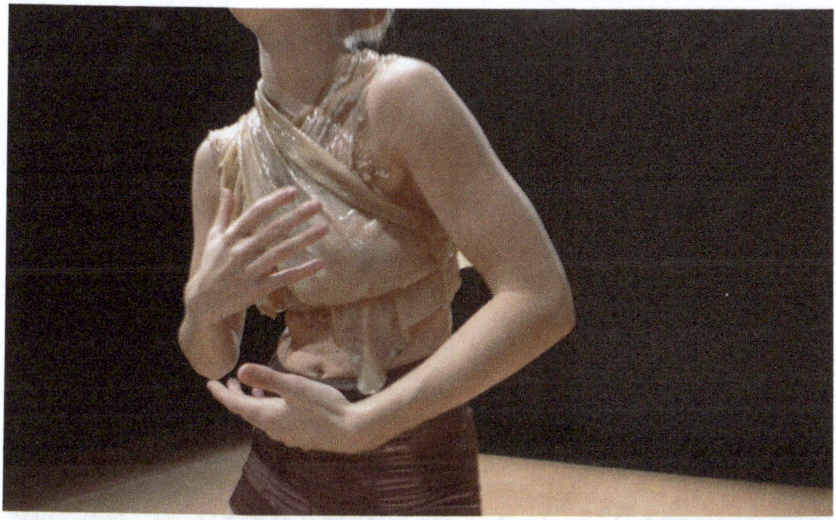

Fig. 3.3 Ingvill Fossheim's bio-costume design for *Posthuman Days* (Jenni-Elina von Bagh, Zodiak—Centre for New Dance, Helsinki, Finland, 2018). (Photo: Ingvill Fossheim)

as a vessel to carry soil and seeds through the streets of São Paulo. For Luanna, costume became the vehicle to form a suspended garden that contained pockets for bags containing plants in which the body becomes both the support and performance space. Both Fossheim and Jimenes used bio-integrated design as a mode of body-nature experimentation and encounters which provoked new performance aesthetics. Taking a more spatial approach, Annike Flo's[17] Ecoscenography work on *cocreat:e:ures* (Vitenparken, Norway, 2018) was a bio-integrated 'trans-species alliance' between the scenographer and the Oyster Mushroom (Flo 2018, 3). In *cocreat:e:ures*, Flo created a fungal installation which became a 'meeting space between humans and non-humans, moving from a human "lab" into a shared space, ending in a fungal space, not accessible to humans' (Ibid.). The scenographic experience of light, moisture and fog was created in direct response to the mushroom's needs and highlighted more-than-human partnerships and experiences (Fig. 3.4).

While in its infancy, there can be little doubt that replacing or merging industrial systems with a biological process is an exciting prospect for

Fig. 3.4 Annike Flo's *cocreat:e:ures* (2018). (Photo: Claudia Lucacel)

theatre practice. Bio-opportunistic design can be a platform to test new strategies and aesthetics for performances that question our human identities and relationship to nature. Its opportunity for transdisciplinary engagement has the potential to radically shift scenographic practices and outcomes. Yet it is also true that bio-opportunistic designers will encounter difficulties in traditional theatre contexts, which often do not have the structures or timelines amenable to experimental work of this nature. Indeed, my own work has encountered these challenges. Collaborating with living materials, ecological and climatic conditions have been a key motivation in projects such as *The Living Stage*, in which I use gardening as a mode of 'making' as well as 'performing' Ecoscenography. However, each *Living Stage* has been conducted outside of the confines of conventional theatre, allowing the freedom to experiment with design approaches that are in tune with living systems, particularly its seasonal rhythms. I return to this topic when describing *The Living Stage* in detail in Chap. 5.

Ecological Design and Regenerative Development

Sim Van der Ryn and Stuart Cowen (2007, x) define ecological design as 'any form of design that minimises environmentally destructive impacts by integrating itself with living processes'. It is a systems-based approach that is ecologically integrated and culturally sensitive. Moving beyond emulating living processes, ecological design embraces broader notions of sustainability, such as conservation, regeneration and stewardship. It implies that design respects, supports and addresses socio-ecological health, including maintaining species diversity, minimising resource depletion and supporting more-than-human communities. Van der Ryn and Cowen propose five ecological design principles that are seen as 'an inspiration to creativity, rather than a definitive set of rules' (xi), which are (1)'solutions grow from place', (2) 'ecological accounting informs design', (3) 'design with nature', (4) 'everyone is a designer' and (5) 'make nature visible'.

With 'everyone is a designer', Van der Ryn and Cowen acknowledge designers as facilitators and catalysts in the cultural processes underlying sustainability which unites ecological, social and cultural values. A tangible example of this ethos is provided by the work of Natalie Jeremijenko. As the director of the Environmental Health Clinic (based at New York University), Jeremijenko blends arts, science and engineering to create radical 'cross-species' artworks that bring poetics to the ecological crisis (Lim 2015). The whimsical clinic—which is part science experiment, part self-help project and part performance—invites 'imPatients' to book an appointment to discuss their prevailing ecological concerns (Schaffer 2008). Rather than a pill or ointment, their prescription comes as a series of actions that are aimed to improve their neighbourhood, encouraging community members to become ecological designers in healing their own places.

Through 'make nature visible', Van der Ryn and Cowen highlight the value of making ecological processes visible to generate a greater awareness of and connection with nature—one that explicitly celebrates the symbiotic relationship across place, ecology, culture and design. Linda Tegg's *Grasslands* (Melbourne International Arts Festival, Melbourne, 2014) was a prime example of celebrating living systems through an engagement with both phenomenological and ecological aesthetics.[18] For the project, Tegg worked with the University of Melbourne and horticulturist John Delpratt to create a site-specific work on the main steps of the State Library of Victoria. Over 15 months, the team researched the biodiversity of the site and cultivated a temporary landscape of over 15,000 plants, including 60 indigenous species that signified the site's rich historical landscape (Backhouse 2014). Tegg's installation stood opposite the entrance to Melbourne Central shopping arcade, offering a highly visible counteraction for the living public artwork, while also honouring the library's place as a meeting point for social and political gathering. Not only did *Grasslands* become an urban refuge for people to seek out or stumble across and learn about urban biodiversity, it also attracted an array of native insects to the busy city location. At the end of its exhibition, the plants were gifted to the general public and the City of Melbourne's Royal Park to enhance biodiversity across the city. *Grasslands* demonstrates an enactment of ecological design as one that acknowledges humans—as well as their developments, social structures and cultural concerns—as an inherent and inseparable part of ecosystems.

An extension of ecological design, regenerative development is a place-based, community-orientated, collaborative and interdisciplinary approach to design and development, with a focus on positive contribution to social and ecological function and evolution (du Plessis 2012, 18). Encompassing notions of circular, bio-inspired and ecological design, regenerative development is the most comprehensive approach to sustainability, which integrates and reconciles complementary environmental design movements to maximise opportunities. The use of the term 'regenerative' calls for the collective activity of 'enhancing life in all its manifestations—human, other species, ecological systems—through an enduring responsibility of stewardship' (Cole 2012, 1). Thus, regenerative development is concerned with creating opportunities for contributive socio-ecological practices that can ripple into broader possibilities to foster thrive-ability (Pedersen Zari and Jenkin 2010, 3).

In regenerative development, designing in response to ecological challenges and opportunities is informed by local contexts, communities and place-specific aspirations (du Plessis 2012). This ethos is evident in the work of Ron Finley[19] (a self-proclaimed 'gangsta gardener') whose international acclaim comes from his unique ability to engage disadvantaged communities in guerrilla-gardening-food-growing in South Central Los Angeles (Finley 2013). Finley's work has inspired thousands of people across the globe to take up food growing, demanding that everyone has the 'right to garden' by 'picking up a shovel' and transforming barren verges and vacant lots into urban food forests. Finley breaks down white and privileged middle-class stereotypes of urban greening, demonstrating that community-engaged food growing is a universal socio-ecological act for all.

Playa Viva—a unique tourist resort in Mexico—provides another example of regenerative development on a grand scale. The resort not only is designed for natural cooling and energy efficiency, but also integrates landscaping inspired by permaculture principles for drought-tolerance and promotion of pollinators. It includes community and ecological programmes such as organic farming, soil regeneration, schools and health clinics, as well as a recycling facility and a local sea turtle nursery to which resort guests are invited to contribute. *Playa Viva* is ambitious in its plans 'to restore at least 85% of the resort's 200 acres (86 hectares) to coastal forests and wetlands' and aims to return 'mangroves, hardwood trees and a variety of indigenous flora and fauna' to the site (Hes and du Plessis 2014, 113–115). However, the largest contribution of *Playa Viva* is its ability to foster meaningful and reciprocal relationships with the local community and their environment. It is the ripples that occur in extending regenerative principles amongst the participants and their wider networks that make it truly regenerative.

In summary, regenerative development aims to empower communities 'across natural, constructed, economic and cultural systems' to enable the potential for positive legacies to emerge (Hoxie et al. 2012, 77). This 'iterative rather than linear' form of engagement encourages interdisciplinary and participatory processes in generating environmental, economic, social and cultural benefits (Eisenberg and Reed 2003). The methodology brings together professionals and community members in a co-creative process in which designs emerge from a deeper understanding of and connection to place. Through this development, designs are constructed as part of larger systems, allowing complex and mutually beneficial interactions between the built environment, the living world and human inhabitants.

Ecological Design and Regenerative Development in Scenography

In the context of the performing arts, ecological design and regenerative development challenge the scenographer to strive for design processes that are capable of creating 'positive legacies' (Hes and du Plessis 2014, 10) for both communities and the environment. Rather than investigating a theatre or a site 'as a collection of things', a regenerative practitioner examines these things as part of a system of interconnected processes and capacities (Haggard 2002, 25). The practitioner must contemplate the interconnections between ecological, social and economic systems at a variety of scales and seek the possibilities of 'place' (a theatre, a warehouse or a whole landscape) to consider both short-term and long-term potential.

A regenerative approach to theatre production is one that interweaves natural, social and cultural aspects to solicit an active co-operative engagement with the more-than-human world. Current examples mostly exist outside of conventional theatre spaces, where performance makers engage with local communities to create work that brings people together around a social-ecological issue across longer time scales. *Organic Theatre* developed by S.N. Sudheer, in the southern Indian state of Kerala, is one such example. The project 'blends art and agriculture' to engage communities in the environmental crisis (Gopika and Eldhose 2020, 543). As Gopika and Eldhose describe:

> *Organic Theatre* is founded on a realisation of the importance of art and indigenous culture which hold reserves of local knowledge regarding the environmental conditions of a place. Here, theatre functions as a channel that facilitates the inter-relationship between cultural and agricultural practices. It creates new ground for interactions within and among communities. (543)

What makes *Organic Theatre* unique is its merging of organic farming with theatre through drama workshops and agricultural activities with local communities. The project is conducted over four years in multiple phases, in which the sprouting of the seeds and harvests coincides with theatre events which empower Indigenous perspectives and histories, as well as highlighting place-specific social and environmental issues that engage the community over time. Director Sudheer was disturbed by the widespread use of pesticides and diminishing traditional farming practices, both of which were destroying the local ecology and culture, and driving disconnection from the land (Nagarajan 2018). The aspiration for *Organic Theatre* was its capacity to function 'as a channel to facilitate real-time cultural and agricultural productivity, creating a play and organic food respectively' (Gopika and Eldhose 2020, 547). Integrating 'agri-culture' with theatre-culture, images of the work depict a thatched house perched alongside stunning landscape contours, mist and mountain scenery, while performers parade in bio-costumes made from reed mats, coir and dried grass (Nagarajan 2018). As Sudheer puts it, 'I like to define it as agri plus culture, a seamless fusion of agrarian activities that [go] hand in hand with theatre, music and other cultural activities' (Sudheer in Nagarajan 2018).

The integration of theatre and food has been fostering a regenerative vision in many parts of the world. Using theatre to build an awareness of food waste is a focus of Practice Tuckshop's *Recess Time*,[20] a lunch-time theatrical feast (created by Xiao Ting, Sim Xin Yi and Joey Cheng) that regularly takes place at the heart of the Singapore's arts district. The participatory work (which has been running since 2018) invites chefs to salvage unwanted produce in local shops which they then incorporate into their menus. The site of the Practice Tuckshop becomes a social space for connection, communion and discussion, where a "Kaypoh" (King/Queen) documents the conversations that emerge from the lunch (Fig. 3.5). The outside gathering space also features an award-winning multi-level architectural installation, *c o o p* (by DO Agency and Nanyang Polytechnic), which is constructed from strand-woven bamboo. The chefs from *Recess Time* regularly use herbs from *c o o p*'s solar-powered aquaponics garden in their daily menu. *Recess Time* artist Xiao Ting (2021) describes her work as one that fosters 'meaningful relationships to build a healthy ecosystem' to 'empower or motivate people to care, to think differently and take action in their own ways' within their field of influence. This is the very essence of what it means to bring a regenerative approach to theatre production.

Fig. 3.5 *Recess Time* (by Ang Xiao Ting, Sim Xin Yi and Joey Cheng), Practice Tuckshop, Singapore. (Photo: Jasmine Lim)

Clare Patey's and Cathy Wren's production of *Feast* (London International Festival of Theatre, 2003) was another project that explored regenerative practices through food. In collaboration with Rosendale Primary School and the Rosendale Allotments, *Feast* was a year-long project that explored children's relationship to food, sustainability and the outdoors. The project included children taking part in artistic, culinary and gardening workshops (planting, cooking, ceramics, natural fabric dying and drama activities) and family evenings with tastings and discussions that culminated in a performative feast. The experience and regenerative capacity of *Feast* is beautifully expressed by Eleanor Margolies (2003, 91), who writes:

> Over two nights, 400 people enjoyed a meal cooked with ingredients grown on the allotment. Tables were laid over the raised beds, labelled to

> show what once grew beneath the guests' feet; children served soup (from watering cans), ratatouille, strawberries, raspberries and bread that had been baked in ovens constructed on the site. A choir sang a specially-composed grace. Throughout history and in every culture, the turning of the seasons has been marked by communal feasts. Through *Feast*, these city children experienced food, their environment and the seasons in new ways: tasting unfamiliar foods, learning how long it takes a tomato to grow, creating a festival, and witnessing the cycle of growth from seed to plant to flower to seed. A year on, this project has spread throughout the school, establishing roots in every classroom. The school's environmental co-ordinator, Olivia Greenaway, says that the allotment is now seen as 'part of the school'… Children who were once astonished by the dirt on freshly-pulled carrots now relish raw beetroot, encouraging each other to taste new foods…The Rosendale experience shows that nature can be experienced in the heart of a city. A local allotment also creates new connections with the community. Neighbouring plotholders of different ages and backgrounds greet the children, comment on the innovations and offer advice. The allotment has become part of the children's world.

What makes the above projects regenerative are their relationship to place, community and unfolding of socio-ecological connection over time. This requires that projects create conditions necessary for sustained, positive evolution (Mang and Reed 2012, 34). Regenerative development, much like other ecological design practices outlined above, requires a stronger level of engagement than conventional sustainability strategies.

Regenerative practice is the ultimate aim of much of my work as an Ecoscenographer, yet I often fall short of achieving it. Most project timelines do not allow the prospect of regenerative development to emerge unless they are conducted over years. Yet there are still short-term benefits to rejoice in: the sharing of food-as-performance, the joy of celebrating the ecological world with audiences, and the re-gifting of sets and costumes to artists and communities in need. These are just some examples that I explore in my own work in Chap. 5. Nevertheless, for me, regenerative thinking is long-term thinking, with the aspiration to create ripples that are beyond my own contribution. This may take the form of physical artefacts that are donated to much-needed charity programmes, but can also include the more intangible and quantifiable, such as connections, networks and ideas from one project that filter through communities long after the event. Despite the challenges of integrating evolutionary thinking into scenography, I argue that it is precisely this aspect of regenerative development that has the capacity to encourage innovation in scenographic practice—a topic that I will explore further in the forthcoming chapters.

Embracing the Whole of Sustainability

The practices outlined in this chapter—eco-efficiency, circular design, eco-effectiveness, bio-inspired design, bio-integrated design, ecological design and regenerative development—can be seen as a holistic spectrum of engaging with sustainability. This continuum is depicted in Bill Reed's framework for engaging in the whole of sustainability (Mang and Reed 2012), which aims to reconcile opposing and complementary viewpoints, strategies, models and philosophical stances of sustainable practice. As discussed in Chap. 2, a whole systems perspective includes both mechanistic and ecological views, using the tools and insights of both (du Plessis and Cole 2011, 438). Adopting an integrative, comprehensive and inclusive approach acknowledges sustainability as a continually evolving discipline that includes both mechanistic and ecological perspectives. This framework can be seen as an 'evolutionary spiral' or a 'gradual unfolding' of human consciousness, where ecological ideas and practices progressively shift from degenerating to regenerating across multiple scales (Reed 2007, 678) (Fig. 3.6).

A holistic approach to sustainability involves expanding beyond, but also keeping, concepts of eco-efficiency. Here, the designer acknowledges the vital foundational role of eco-efficiency in sustainability and demonstrates that more challenging ecological approaches can be synergistic with eco-efficient strategies while advancing towards regenerative development. Sustainability is seen as transitional, experimental and progressive, where creative potential becomes one of shifting from limitation to one of opportunity. This philosophy remains at the crux of transforming ecological *thinking* (Chap. 2) into ecological *practice* in Ecoscenography.

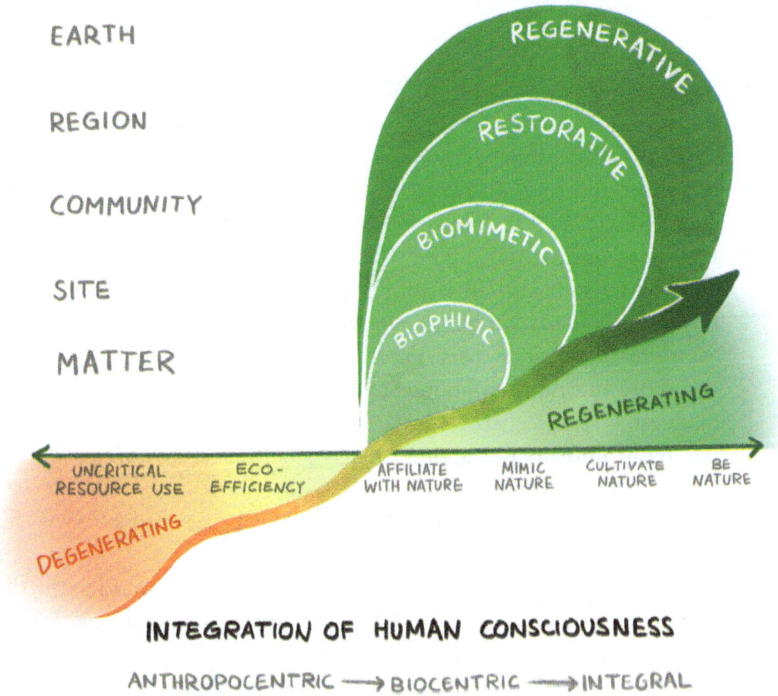

Fig. 3.6 An interpretation of Bill Reed's framework for the whole of sustainability (Regenesis 2000–2011). (Illustration by Gisela Beer)

Embracing a holistic and expanded approach to sustainability highlights new opportunities for ecological practice that are embraced through a lens of possibility. Even if the more ambitious ecological design practices (such as regenerative development) can only be primarily investigated outside of existing paradigms, I propose that the performing arts requires a vision that it can aspire to—one that includes the potential for creative engagement and demonstrates an inspiring pathway for embracing sustainability. The next chapter presents Ecoscenography's cyclical framework as a guide to creating ecologically conscious, fun, inspiring and provocative spatial experiences. It delves into the ways in which to harness the potential of Ecoscenography as it applies to the specificity of theatre production across a range of platforms.

Notes

1. More information about the Willow Theatre can be found via: https://www.archdaily.com/440635/willow-theater-tim-lai-architect-brad-steinmetz-stage-design.
2. For more information about *À Ố Làng Phố*, see: https://theplusones.com/sydney/2019/06/13/vietnamese-circus-opera-house/.
3. For more information about *Glimpsing Air Pockets*, see: https://www.monakastell.com/2019/01/30/glimpsing-air-pockets/.
4. For more information about Ahrend, see: https://www.ahrend.com/en/solutions/furniture-as-a-service.
5. For more information about Street Sleeper, see: https://www.upcycle-that.com/street-sleeper/.
6. For more information about Ananas Anam, see: https://www.ananas-anam.com/about-us/.
7. For more information about Chip[s] Board, see: https://www.chips-board.com/.
8. *Un reseau Translucide* received an 'Honorary Mention' award in the PRIX ARS ELECTRONICA Hybrid Art category (2012). For more information, see: https://www.pruelang.com/2011/un-reseau-translucide/.
9. For more information about *vox:lumen*, see: http://www.ianpgarrett.com/portfolio/vox-lumen/.
10. For more information about *Plastivore*, see: https://www.smh.com.au/environment/sustainability/the-art-of-clean-living-20190722-p529jz.html.
11. For more information about Roya Aghighi's work, see: https://www.roya-aghighi.com/home.html.
12. For more information about Elissa Brunato's work, see: https://www.elissabrunato.com.
13. For more information about the *Growing Pavilion*, see: https://thegrowingpavilion.com.
14. For more information about Martijn Straatman's work, see: https://studiotinus.com/collections.
15. For more information about Aniela Hoitink's work, see: https://neffa.nl.
16. For more information about Luana Jimenes' work, see: https://emperformance.wixsite.com/nucleodeperformance/jardim-suspenso
17. For more information about Annike Flo's work, see: http://www.annike flo.com/cocreateures.

18. For more information on *Grasslands* and Linda Tegg's work, see: http://www.lindategg.com/grasslands.html.
19. For more information about Ron Finley's work, see: https://www.ted.com/talks/ron_finley_a_guerrilla_gardener_in_south_central_la.
20. For more information about *Recess Time*, see: https://www.viddsee.com/video/f0od-and-art/71pwo?locale=en.

References

Asia TOPA. 2020. Creating the World's First Bamboo Circus with Director Tuấn Lê. *The Guardian*. https://www.theguardian.com/asia-topa-2020-program/2020/jan/23/creating-the-worlds-first-bamboo-circus-with-director-tuan-le.

Backhouse, Megan. 2014. Grass From the Past in Swanston Street. *The Sydney Morning Herald*. https://www.smh.com.au/entertainment/art-and-design/grass-from-the-past-in-swanston-street-20141010-1146k4.html.

Benyus, Janine. 2002. *Biomimicry: Innovation Inspired By Nature*. New York, USA: HarperCollins.

Birkeland, Janis. 2012. Design Blindness in Sustainable Development: From Closed to Open Systems Design Thinking. *Journal of Urban Design* 17 (2): 163–187.

Block, India. 2019a. Bio Iridescent Sequin Made From Trees Responds to Growing Demand for Greener Fashion. *Dezeen*. https://www.dezeen.com/2019/08/02/bio-iridescent-sequin-elissa-brunato-sustainable-fashion/.

———. 2019b. Biogarmentry Clothes Can Photosynthesise Like Plants. *Dezeen*. https://www.dezeen.com/2019/10/02/biogarmentry-roya-aghighi-living-clothes-photosynthesis/.

Brundtland Commission. 1987. Our Common Future. In *World Commission on Environment and Development*, ed. Volker Hauff. http://www.un-documents.net/wced-ocf.htm.

Brunner, Paul, and Michael Mehler. 2013. Theatre Design and Production Reimagined: Four Principles for a Sustainable Future. *Theatre Design and Technology* 49 (3): 23–32.

Carson, Rachel. 1962. *Silent Spring*. Boston, USA: Houghton Mifflin Company.

Charter, Martin. 2018. *Designing for the Circular Economy*. New York, USA: Routledge.

Chow, Denise. 2018. Eco-friendly 'Algae Curtains' Could Help Curb Air Pollution in Crowded Cities: Draped Over Buildings, the Living Walls Scrub Carbon Dioxide From the Air While Pumping Out Oxygen. *NBN News*. https://www.nbcnews.com/mach/science/eco-friendly-algae-curtains-could-help-curb-air-pollution-crowded-ncna946946.

Cole, Raymond. 2012. Regenerative Design and Development: Current Theory and Practice. *Building Research and Information* 40 (1): 1–6.

Cooper, Megan. 2020. This Designer Created a Bookshelf That Can Be Turned Into a Coffin After Death. *The Modern Met*. https://mymodernmet.com/william-warren-shelves-of-life/.

DeSimone, Livio, and Frank Popoff. 1997. *Eco-efficiency: The Business Link to Sustainable Development*. Cambridge, USA: MIT Press.

du Plessis, Chrisna. 2012. Towards a Regenerative Paradigm for the Built Environment. *Building Research and Information* 40 (1): 7–22.

du Plessis, Chrisna, and Raymond Cole. 2011. Motivating Change: Shifting the Paradigm. *Building Research and Information* 39 (5): 436–449.

Earley, Rebecca. 2011. *Worn Again: Rethinking Recycled Textiles (2005–2009)*. London, UK: Arts and Humanities Research Council.

Eisenberg, David, and Bill Reed. 2003. *Regenerative Design: Toward the Re-Integration of Human Systems within Nature*. Paper presented at the Greenbuild Conference, Pittsburgh, USA.

Finley, Ron. 2013. A Guerrilla Gardener in South Central LA. *TED*: https://www.ted.com/talks/ron_finley_a_guerrilla_gardener_in_south_central_la?language=en.

Flo, Aninike. 2018. *Cocreat:e:ures*. Masters Thesis, Norwegian Theatre Academy, http://www.annikeflo.com/cocreateures.

Fosseheim, Ingvill. 2019. *Costume Matter: Exploring Microorganisms, Fungi and Berries as Biobased Material in Contemporary Costume Design*. Masters Thesis, Aalto University School of Arts, Design and Architecture, https://issuu.com/fossheim.ingvill/docs/fossheim_costumematter_webversion.

Garrett, Ian. 2016. vox:lumen. https://www.ianpgarrett.com/portfolio/vox-lumen/.

Goldmark, Sandra. 2020. *Fixation: How to Have Stuff Without Breaking the Planet*. Washington, DC: Island Press.

Gopika, Selvakumar, and Adakkaravayalil Yoyakky Eldhose. 2020. Experiments in Organic Theatre. *Research in Drama Education: The Journal of Applied Theatre and Performance* 25 (4): 543–547.

Graedel, Thomas, Paul Reaves Comrie, and Janine Sekutowski. 1995. Green Product Design. *T&T Technical Journal* 74 (6): 17–25.

Haggard, Ben. 2002. Green to the Power of Three. *Environmental Design and Construction* 5 (2): 24–31.

Hes, Dominique, and Chrisna du Plessis. 2014. *Designing for Hope: Pathways to Regenerative Sustainability*. Hoboken, USA: Taylor and Francis.

Hoxie, Christina, Robert Berkebile, and Joel Todd. 2012. Stimulating Regenerative Development through Community Dialogue. *Building Research and Information* 40 (1): 65–80.

Kastell, Mona, and Hannah Myers. 2019. Ecoscenography in Action: Bridging Stage Design with Nature Connection to Shape Sustainable Communities and Wellbeing. *Scene* 7 (1–2): 29–44.

Kellert, Stephen, Judith Heerwagen, and Martin Mador. 2008. *Biophilic Design: The Theory, Science and Practice of Bringing Buildings to Life*. Hoboken, USA: John Wiley & Sons.

Lang, Prue. 2020. Un réseau translucide. https://www.pruelang.com/2011/un-reseau-translucide/.

Leonardi, Thierry. 2020. Opera Production and the Circular Economy: Interview with Thierry Leonardi (Lyon Opera). In *Ecoscenography: Adventures in a New Paradigm for Performance Making*, ed. Tanja Beer. https://ecoscenography.com/2020/11/24/opera-production-the-circular-economy-interview-with thierry-leonardi-lyon-opera/.

Lewis-Fitzgerald, Erin. 2020. *Modern Mending: From Rags to Riches, How to Reduce Waste and Raise Your Style*. Melbourne, Australia: Affirm Press.

Lim, Eugenia. 2015. The Science of Citizens: Natalie Jeremijenko. In *Assemble Papers: The Culture of Living Closer Together*. https://assemblepapers.com.au/2015/03/17/the-science-of-citizens-natalie-jeremijenko/.

Mang, Pamela, and Ben Haggard. 2016. *Regenerative Development and Design: A Framework for Evolving Sustainability*. Hoboken, USA: John Wiley and Sons.

Mang, Pamela, and Bill Reed. 2012. Designing from Place: A Regenerative Framework and Methodology. *Building Research and Information* 40 (1): 23–38.

Margolies, Eleanor. 2003. The Allotment with Roots in Every Classroom. In *Landing Stages: Selection from the Ashden Directory*, ed. Wallace Heim and Eleanor Margolies, 90–92. London, UK: Crinkle Crankle Press.

McDonough, William, and Michael Braungart. 2002. *Cradle to Cradle: Remaking the Way We Make Things*. New York, USA: North Point Press.

———. 2013. *The Upcycle*. New York, USA: North Point Press.

Morris, Ali. 2018. MIT Engineers Transform Plants into Lights. *Dezeen*. https://www.dezeen.com/2018/02/01/movie-mit-engineers-firefly-enzyme-glowing-plants-video/.

Myers, William. 2012. *Bio Design: Nature, Science, Creativity*. London, UK: Thames and Hudson.

Nagarajan, Saraswathy. 2018. Organic Theatre, Conceptualised by SN Sudheer, is Breaking New Ground in Kerala. *The Hindu*. https://www.thehindu.com/entertainment/theatre/organic-theatre-as-conceptualised-by-sn-sudheer-is-breaking-new-ground-in-kerala-with-its-emphasis-on-reviving-ties-between-farming-and-culture/article25622206.ece.

Nordhaus, Ted, and Michael Shellenberger. 2009. *Break Through: From the Death of Environmentalism to the Politics of Responsibility*. Oregon, USA: Blackstone Audio Inc.

Northover, Kylie. 2019. The Art of Clean Living. *The Sydney Morning Herald*. https://www.smh.com.au/environment/sustainability/the-art-of-clean-living-20190722-p529jz.html.

Pallister, James. 2014. Glowing Trees Could Be Used "Instead of Street Lighting" says Daan Roosegaarde. *Dezeen*. https://www.dezeen.com/2014/03/24/movie-sxsw-daan-roosegarde-glow-in-dark-trees/.

Pedersen Zari, Maibritt. 2015. Ecosystem Services Analysis: Mimicking Ecosystem Services for Regenerative Urban Design. *International Journal of Sustainable Built Environment* 4 (1): 145–157.

———. 2016. Mimicking Ecosystems for Bio-Inspired Regenerative Built Environments. *Journal of Intelligent Buildings International* 8 (2): 57–77.

Pedersen Zari, Maibritt, and Sarah Jenkin. 2010. Re-defining Cutting Edge Sustainable Design: From Eco-efficiency to Regenerative Development. Paper presented at the Sustainable Building Conference Wellington, New Zealand.

Pownall, Augusta. 2019. Pavilion Grown from Mycelium Acts as Pop-up Performance Space at Dutch Design Week. *Dezeen*. https://www.dezeen.com/2019/10/29/growing-pavilion-mycelium-dutch-design-week/?li_source=LI&li_medium=rhs_block_3.

Reed, Bill. 2007. Forum Shifting from 'Sustainability' to Regeneration. *Building Research and Information* 35 (6): 674–680.

Robinson, John, and Raymond Cole. 2015. Theoretical Underpinnings of Regenerative Sustainability. *Building Research and Information* 43 (2): 133–143.

Schaffer, Amanda. 2008. Prescriptions for Health, the Environmental Kind. *The New York Times*. https://www.nytimes.com/2008/08/12/health/12clin.html.
St. Pierre, Louise. 2019. Design and Nature: A History. In *Design and Nature: A Partnership*, eds. Kate Fletcher, Louise St. Pierre, and Mathilda Tham. London, UK: Routledge.
Ting, Xiao. 2021. Community Engaged Eco-theatre Action: Interview with Xiao Ting (Singapore). In *Ecoscenography: Adventures in a New Paradigm for Performance Making*, ed. Tanja Beer. https://ecoscenography.com/2021/01/25/community-engaged-eco-theatre-action-interview-with-xiao-ting-singapore/.
Van der Ryn, Sim, and Stuart Cowan. 1996. *Ecological Design*. Washington, DC: Island Press.
———. 2007. *Ecological Design: Tenth Anniversary Edition*. 10th anniversary ed. Washington, DC: Island Press.
Watson, Julia. 2019. *Lo-TEK: Design by Radical Indigenism*. Cologne, Germany: Taschen.
Webster, Ken. 2017. *The Circular Economy: A Wealth of Flows—2nd Edition*. Ellen MacArthur Foundation. https://www.ellenmacarthurfoundation.org/publications/the-circular-economy-a-wealth-of-flows-2nd-edition.
Wilson, Edward O. 1984. *Biophilia*. Cambridge, USA: Harvard University Press.
Woynarski, Lisa. 2020. *Ecodramaturgies: Theatre, Performance and Climate Change*. Cham, Switzerland: Palgrave Macmillan.

4

Ecoscenography in Three Acts

T. Beer, *Ecoscenography*, https://doi.org/10.1007/978-981-16-7178-4_4

My premise for Ecoscenography is to reconsider performance design practices in a way that looks to the living world for inspiration. To cogitate how theatre production can draw upon time-tested and cyclic patterns found in nature, and ways of working that are inspired by 3.8 billion years of creating beneficial life-enhancing conditions (Pedersen Zari 2016; Benyus 2002). Encompassing the complex entanglement of material processes, relationships and effects—both seen and unseen—is at the core of bringing an ecological ethic to theatre production. Ecoscenography augments the 'liveness' of the planet by relating and unifying performance design with the wider system of ecological organisation. It is intended as 'relational' in the sense that it is 'co-constituted by spatiotemporal and sociopolitical values' (Rawes 2013, 1). This means viewing design from the perspective of relationships, rather than separate entities, as a way of making sense of the extensive impacts of interactive material-spatial processes. Embracing these 'relational materialities' is an essential part of starting to disrupt the linear ethos that has prevailed in our industry for so long.

Relationality is fundamental to nature's processes which have developed over billions of years of co-evolution. Bacteria, fungi, plants and animals co-create together to circulate nutrients and develop healthy and safe metabolisms in which any form of 'waste' is treated as a valuable resource for further transformation. According to the Law of Conservation of Matter, nothing is ever created nor destroyed; it only transforms from one form to another. For example, water circulates between the air, oceans and living things, taking on different compositions and forms. Nature's collaborators—soil, bacteria, fungi, plants and animals—are always in a state of transition, co-adapting and evolving. Importantly, living systems do not value one stage of being over the other; instead, they are part of a process of 'further becoming' in which each stage is crucial and valued:

> A seed dropped into the soil grows into a plant, produces leaves, and flowers. The flowers bear seeds, which drop to the ground and become new plants, and so the cycle continues. The cycle of life is a view of existence in which all things end at the beginning in a never-ending cycle. While the

> plant lives, it thrives on the rain that falls from the sky. It takes in moisture and nutrients from the soil through its roots, and cycles the moisture back to the sky by breathing it out through its leaves, in a process called transpiration. (Nelson Education Ltd 2004, 86)

As discussed in Chap. 3, cyclic ways of thinking and doing are a crucial part of transitioning from a mechanistic (linear, reductionist, machine-like) worldview to an ecological (holistic, relational, interdependent) worldview. Cycles are evident not only in ecology but also in cultural traditions of understanding universal processes of 'creation, embodiment, preservation, destruction, and release' (Hes and du Plessis 2014, 29). Indigenous cultures have long nurtured the value of cycles as a way of bringing human practices and ways of living closer to the earth, not only as a necessary part of survival, but also in seeking prosperity and abundance. First Peoples have used cyclic and transitional events in the natural world to predict seasonal changes that have been an important and necessary part of their ecological and cultural practices (Kimmerer 2013; Pascoe 2014).

At the heart of Ecoscenography is the notion of integrating ecological thinking into all stages of theatre production, which means to give value to the whole process. Foundationally, this is about realigning ourselves with the cyclicity of living systems, where the artistic 'outcome' is one of 'further becoming' that extends well beyond Opening Night. This chapter introduces Ecoscenography's cyclical framework as a guiding structure for best practice. It builds on the previous chapters to demonstrate how ecological thinking and practice can be specifically applied to the context of the performing arts, fostering new practices and aesthetics that enhance the social and environmental advocacy of our field. This proposal is by no means a panacea, but a starting point—a testing ground that is informed by the ecological theories and approaches presented in this book. It is a regenerative reimagining of the future of performance design.

From Linear to Cyclical Ways of Thinking and Doing

If we are to fully embrace ecological ways of thinking and doing, a key consideration must be how vitality and agency can be attributed to all stages of theatre production. This entails moving our focus beyond the performance season, to one that also acknowledges and celebrates the 'making' and 'un-making' of our practices. Unlike typical theatre productions where the performance season is precedent, Ecoscenography shifts our focus from the performance season, to also consider the 'making of' and 'end-of-performance-life' stages of design as part of the aesthetic process. This entails reframing scenography beyond the significance of the performance season, where the 'behind the scenes' components of theatrical design (preparation, construction and distribution) are brought 'into the spotlight' as a valuable part of the aesthetic experience.

Ecoscenography comprises three 'stages' or 'acts' that are considered equally fundamental to the aesthetic consideration of the work—co-creation (pre-production), celebration (production) and circulation (post-production). 'Co-creation' acknowledges the highly collaborative phase of theatre making as one that is co-extensive and co-productive, where ecological thinking is aligned with the creative process of scenography; 'celebration' embraces the stage as a platform to bring people together to showcase ecological ideas and practices, and 'circulation' brings value to ecological design processes continuing beyond the theatre—for materials and resources to be distributed out into the wider community—for further making and transformation. Inspired by ecological processes, the three stages of Ecoscenography are cyclic, where 'circulation' (the gifting of ideas and resources) leads back to 'co-creation' (new materials, ideas and collaborations fuelling the artistic process), allowing the cycle of production to continue spiralling forward. In this context, the performance season is seen as an active celebration along a design's journey of further becoming. The loop-sequencing of 'co-creation—celebration—circulation' helps expand ecological understanding, perception and action, which deepens with each iteration of practice.

Ecoscenography is biomimetic or bio-inspired in that it draws inspiration from living systems to stimulate patterns and processes based on life-enhancing conditions (Benyus 2002; Mang and Reed 2012). For example, the life-cycle of plant can be seen as a process of 'co-creation' (germination, growth) that leads to 'celebration' (flowering, pollination, fruiting) and 'circulation' (seed dispersal, consumption by animals, and/or decomposition and nutrient recycling), which in turn leads back to 'co-creation' (germination, growth). By redistributing value across all stages of the artistic process, the 'three C's' of co-creation—celebration—circulation also provides an opportunity to engage audiences and communities in the making and sharing of eco-creative processes inspired by nature's systems, both in and beyond the production. While theatre artists may utilise all three stages in the creation of their work, this chapter uses the emphasis of each category to draw out the specific characteristics of each process as a way of gaining a deeper knowledge of how these approaches, techniques and tools manifest notions of Ecoscenography.

Act 1: Co-creation

Co-creation is a necessary part of all living systems. A thriving ecosystem is one that is not only resourceful and contributive but also adaptive to local collaborators and conditions. For example, plants rely on the sharing of resources to function, such as adequate space, water, nutrients and sunlight, whilst also adapting to local conditions and giving back to the system by emitting oxygen, providing habitat, shelter, shade and food. Just as a plant co-creates with living systems in order to thrive, Ecoscenography is best achieved when designers are in tune with the serendipitous opportunities and resources around them, with the impetus to collaborate, share ideas and resources that can contribute back to the system.

Ecoscenography is a co-creative act, where more-than-human actants are co-entangled and co-complicit in each other's making. From an ecological perspective, nothing exists by itself; we are always interacting and co-creating with our co-extensive selves. The notion of 'worlding' is highlighted by ecofeminist writer Donna Haraway (2016), as one which exemplifies the world's 'ongoingness' across entangled human-world relations (e.g. abiotic, biotic and technotic). As Haraway (2016, 58) describes:

> Nothing makes itself; nothing is really autopoietic or self organizing... *sympoiesis* is a word proper to complex, dynamic, responsive, situated, historical systems...it is a word for worlding-with, in company.

I suggest that the notion of 'worlding-with' is a powerful and much-needed proposition for the merging of ecology and co-creativity in scenography. 'Worlding-with' embraces and expands acts of worlding or staging in scenography (Hann 2018) to give focus to our co-extensive practices. 'Worlding-with' or 'staging-with' is to bring about a mutually beneficial partnership with the more-than-human world. As Haraway (2016, 63) contends, it is what ties 'together human and nonhuman ecologies, evolution, development, story, effects, performances', thereby dissolving perceived binaries between artists, places, materials, audiences and the broader ecosystem.

Co-creation is one of partnering with place, where Ecoscenographic opportunity is found in uncovering the abundance and complexity of more-than-human possibilities (McKinney 2015, 91–92). Embracing place-specific contexts and possibilities invites scenographers to access local knowledge and insights, opening up new avenues for creative exploration. As ecological design experts Van der Ryn and Cowan (2007) assert, much of our resourcefulness comes from being receptive to what is locally available in the places in which we find ourselves. Allowing space for this porosity not only has the capacity to minimise environmental impacts through the accessing and sharing of resources, but also can stimulate new partnerships, audiences and networks.

Even within the confines of the theatre building, the notion of co-creating with 'place' is an effective strategy that includes accessing readily available features, resources and materials. Responding to 'place' inside the theatre may involve making use of existing theatre walls, lighting rigs or existing stock as a way of maximising both economic and environmental efficiencies. As Goldmark asserts, 'Using things that already exist—whether in stock, or in rental houses, or purchased used, is one of the simplest ways to reduce the environmental impact of a design' (2017, 23). Working with what is already available does not need to be seen as a limitation, but rather as a way of accessing vital information for scenographic possibilities. Even within touring contexts, a theatre artist can consider co-creating with place by substituting generic set elements (rostra, truss, lighting rigs) with localised resources or collaborating with local performers and designers. Essentially, Ecoscenography begins with responding to where we are, here and now, as part of the co-creation process.

Noémie Avidar (2020) defines her own Ecoscenographic practice as 'shaping a space for a performance in a way that acknowledges and embraces its surroundings'. Place-based responses were an integral part of Avidar's set design for *Winslow* (L'Escaouette, Moncton, New Brunswick, 2019), which told the story of the French-Canadian deportation events of 1755, through the perspective of Lieutenant-Colonel John Winslow, commander of the Anglo-American troops. For Avidar, it was through relating the historic incidents of the play to the local fishing community and the economic, social and historical connection of the Acadian population to the sea that prompted her to seek out opportunities to use site-specific materials. This included making use of the overabundance of discarded fishing ropes that became the central wall feature for the set design (Fig. 4.1). Constructed from 1462 reclaimed ropes, the stunning backdrop not only created a textured projection surface, it also authentically supported the setting and themes of the play (Fig. 4.2). For Avidar, 'engaging authentically with people or place as part of the aesthetic process' stimulates a way of thinking that is emergent and receptive:

> It is about extending the idea of a public performance to an everyday performance, a kind of game. I get to meet so many people and to acknowl-

Fig. 4.1 Noémie Avidar's array of reclaimed ropes sourced from the local fishing community. (Photo: Noémie Avidar)

> edge their presence in my creative process is a pure joy…It is about taking into account the history, the integrity and ability of each of these elements so they play a role in the theatrical piece. (Ibid.)

The joy, adventure and satisfaction that comes from designing in collaboration with people and place is also reiterated by Ruth Stringer[1] (2021), who credits her formative experience in site-specific theatre as one that has allowed her to see place differently and to respond to its potential:

> Ever since my first role with the National Theatre Wales, I have looked at what the landscape provides, allowing it to inspire and build my designs. I remember looking out of the window of our facility one morning and seeing a shopping trolley half-buried in a sand dune. I knew it would be perfect for a large, mobile torch that I needed to make, and loved the idea

Fig. 4.2 Noémie Avidar's projection screen (made from 1462 reclaimed fishing ropes) for *Winslow* (L'Escaouette, Moncton, New Brunswick, 2019). (Photo: Noémie Avidar)

> of transforming this forgotten piece of rubbish into a prop! I made a light installation from old abandoned umbrellas, which I sourced from pubs and nightclubs. I was inundated with items that had been long-forgotten, waiting to be turned into art. Site-specific theatre really opened up my practice in terms of responding to a site, and working with it, rather than imposing my vision upon it, which I think is fundamental to Ecoscenography.

Stringer's practice (Fig. 4.3) demonstrates how designing for site-specific performance can foreground ecological processes. In site-specific theatre, it is the existing conditions and opportunities that take centre stage:

> There is little attempt at scenic illusion, at the substitution of theatrical properties and devices constructed in replica forms; rather, fragments of

Fig. 4.3 Ruth Stringer's costume design made from plastic waste (Egin Residency, National Theatre of Wales, 2019). (Photo: Ruth Stringer)

> the real world—real things—are selected, enter into and are reconfigured in unfamiliar juxtapositions of like and unlike, in assemblages unique to each production, albeit within the selective metonym of performance. (Pearson 2018, 296)

From a site-specific perspective, 'real things' take priority over 'constructed replicas' to authenticate place in performance. Here, found objects and local resources (often in their most genuine and unaltered form) become creative solutions that also maximise sustainability outcomes.

The 'scenographer as gleaner' is a key strategy for pursuing place-based sustainability responses. Ecoscenographers forage across streetscapes and abandoned festival sites, and scavenge tip shops, recycle centres and charity shops. Here, extending the use of materials is not necessarily approached out of austerity but fuelled by a desire for invention and ingenuity—a way of rethinking design in response to ecological values. A notable example is Tony award-winning Broadway set designer and sustainability advocate Donyale Werle,[2] who works almost exclusively with salvaged materials, and describes her process as literally rummaging through local garbage sites to find treasures and ideas for her designs that fuel her creative process (Kompanek 2012, 30). Werle's large-scale set design for *Peter and the Starcatcher* (2012) was almost entirely made from recycled, borrowed, donated or salvaged materials, including costume fabrics from past productions and discarded household materials sourced from local junkyards, charity shops and recycled art-material suppliers. Similarly, Ecoscenographer Andrea Carr[3] has created stunning designs from found resources, including transforming dozens of camping supplies abandoned at a local music festival into striking chrysalis-shaped costumes for *Stuck* (HOAX Theatre, London, 2017) (Fig. 4.4). For Werle and Carr, the economic poverty or stigma of material debris has in no way hindered the rich and varied performance spaces they have created.

Importantly, Ecoscenography not only accentuates opportunities for designers to engage with local materials but also encourages theatre makers to consider how communities might become part of the co-creation processes. This approach can foster new networks, audiences, skills and possibilities that strengthen the meaning and outreach of the project as well as providing a platform to foster equity and togetherness on global-to-local issues. Such community-engaged processes were a focus of *Portage* (Arts House, Melbourne, 2016–2020), a multi-platform climate art project where audiences were engaged across four stages (*Raft, Flotilla, Shelter, Camp*). A collaboration between artist Jen Rae and designers Mittul Vahanvati and Munir Vahanvati (Giant Grass), *Portage* engaged local participants in free bamboo-construction workshops led by six intercultural master weavers from different Indigenous backgrounds to create watercraft for an immersive performance installation. A key part of the project was to unearth 'overlooked skills, knowledges and values that

Fig. 4.4 Andrea Carr's set and costume design for *Stuck* featured in Aesthetic Art Prize Anthology, Future Now 2017 (HOAX Theatre, London, 2017). (Photo: Kasia Rucińska)

might offer salvation in the years ahead…a call to mobilise, collaborate and arrive at purpose' in response to the climate emergency (Fair Share Fare 2019).

Community crafting as Ecoscenography is also evident in the work of Mexican scenographer Aris Pretelin-Estéves,[4] who sees weaving as a restorative activity that stimulates social connections and reinforms participants' relationships to their local environment. Similar to *Portage*, Pretelin-Estéves' ongoing work (entitled *Tejidos*) consists of three steps in which a collectively woven artistic piece made with recycled material is created and installed. This includes (1) *Memorias Tejidas*, a collective weaving workshop to generate the material; (2) *Tejedoras de Caminos*, a scenographic performance where designers and the public weave the material across trees (Fig. 4.5); and (3) *Atlas de Memorias*, an exhibition of the work through video documentation. Pretelin-Estéves (2021) literally weaves social and ecological components together in urban green

Fig. 4.5 *Tejidos* activation (Aris Pretelin-Estéves), Prague Quadrennial, 2019. (Photo: Carlos Casasola)

spaces as a way of reconciling intercultural human-nature communities both practically and symbolically. As she explains:

> TEJIDOS is a project created with, for and by the community. It involves several scenographic actions that seek to transform urban green spaces into inclusive gathering places and aims to reverse the neglect and devastation that surrounds us. TEJIDOS begins with the donation and recycling of discarded garments that participants later transform, weave, design and install in a green area of the city. The fabric acts as a guiding axis for the project and is a metaphor for the multiple networks that connects us to the environment and the broader world. The scenography for TEJIDOS is created collectively, and participants undertake a journey during which the green space and its surroundings are perceived from different points of view, unveiling the history-memory of the space and therefore, its natural

> liveliness. In creating a collective, tactile, and memorable experience, a convivial bond is woven, fostering a sense of care for our green spaces. Knitting and weaving are simple somatic actions that, by their continuous repetition, relaxes the body and encourages participants to listen and engage in dialogue with one another. These conversations and ideas are subsequently shared with institutions, committees, and neighbourhood associations to generate agreements and actions that will transform and take care of green spaces in the long term. TEJIDOS responds to a 'Povera' aesthetic, seeking the transformation of waste materials by exposing them to a natural environment that modifies and activates them and creates a sense of meaning. The richness of the project lies in its ability to promote 'encounter-spaces' for collective action—to regenerate the social fabric and to resist the social-environmental crisis that we are experiencing right now in Mexico. (2021)

The strength of *Portage* and *Tejidos* is in their ability to make eco-creative processes and aesthetics visible and accessible, breaking down perceived binaries between artists and the wider public. Coming together to 'make something happen' is the very essence of capacity-building on ecological issues. This corresponds with Ecoscenography's regenerative ambition to directly engage communities in experiences of creative sustainability that foster tangible opportunities of hope, abundance and regeneration.

To summarise, co-creation in Ecoscenography is a form of engagement where theatre makers align themselves with place—to open up to the 'worlding' of collaboration that takes into account co-extensive histories and futures across venues, communities and materials. In this process of 'making with', the scenographer is no longer seen as a 'space manipulator' but instead becomes a mediator, facilitator and co-creator. Here, the opportunities lie in the scenographer's ability to activate socio-ecological potential by seeking out opportunities for sharing, collaborating and engaging with others in the pursuit of a more integrated and rewarding creative practice.

Act 2: Celebration

'Celebration' in nature can be witnessed through the extraordinary show that many plants perform through the act of flowering, thereby attracting a diversity of insects and pollinators. The blooming of plants not only highlights moments of beauty and connection, their flowers also serve a very important ecological purpose in the fertilisation and propagation of species viability. Just as nature embraces flowering, pollination and fruiting along its journey of further becoming, Ecoscenography nurtures the performance season not as an endpoint, but as a vibrant and highly visible celebratory mid-point in the

creative process. We celebrate the emergence of the show, and we celebrate the stage as a platform for the meaningful communion, connection and cross-pollination of culture and ideas.

The etymological roots of the English word 'celebrate' are derived from the Latin *celebrare*, meaning 'to assemble to honour' or to 'observe, respect, perform, commemorate' together. The notion of 'celebration' in Ecoscenography is one of 'gathering' to observe, respect and perform ecological practices in real time and place—one that links our longstanding cultural practices (rituals, ceremonies, rites and traditions) with sustainability.

McKinney and Palmer (2017, 5–6, 8) highlight the capacity for scenography to create 'empathetic encounters' or tactics in which 'participants are enabled to see familiar environments in fresh ways'. Creating 'empathetic encounters' is essential to 'celebration' in Ecoscenography, where the stage becomes a platform for empowering and catalysing ecological potential across audiences. The very act of making and performing together provides opportunities for collective memories which can act as powerful agents of change (Mang and Reed 2012, 29–30), permeating psycho-social spheres and influencing the lives of those connected to them (Bal and van Boheemen 2009). Theatre provides a rich platform for performing experiences and narratives that can strengthen our common imagination, germinating new ways of collectively thinking and being in the world. As eco-playwright Teresa J. May (2013, 193) writes, 'stories create a matrix of belonging, a living tissue between past and present, and between human and non-human communities'. The idea of coming together to witness ecological ideas and practices is important; it helps to strengthen community bonds and reinforces the concept that sustainability is something that is collectively valued (Robertson 2014, 297). As Fried and May (1992) state:

> Historically theatre brought people together to witness enactments in different ways of looking at the world. Out of these cosmologies arose the culture's values, ethics, priorities, methodologies…its patterns of thinking and doing…Theatre can be a hothouse for germinating new ideas and imagining new possibilities. It can be a collective dreaming through which we re-envision what it is to be human. (Fried and May 1992, v)

Critiquing cultural values and imagining new possibilities prompted director Katie Mitchell and designer Chloe Lamford to seek out novel, sustainable approaches for *Atmen,* a German version of Duncan Macmillan's climate change play *Lungs* at Berlin's Schaubühne (2013). The team worked with Electric Pedals to stage the entire performance using human-generated power which was visibly controlled by the actors on stage. The show was focused on two protagonists who pedalled their way through topics such as overpopulation, climate change and species extinction, while four supporting cyclists helped power lights, sound and projection from the sidelines. As audience member Summer (Banks 2014) writes:

> The electricity production was always visible, as soon as the actors slowed down, took a break for water, or toweled off their brows, the red energy meters on the front of their podiums clicked down and began to blink, then the lights would fade until they picked up the pace again…And as the performance went on, the human-power component continues to shape and inform the text. Lower energy from the actors translates nearly instantaneously into lower lights, flickering and sputtering. When the male protagonist passes away at the show's end, he stops cycling and his lights slowly fade out. A simple yet profound image, with plenty of room for interpretation.

Part of the novelty of *Atmen* was the way in which sustainability was made uncompromisingly transparent as part of the experience, one where Ecoscenography becomes a highly visible dramaturgical approach to production.

Nevertheless, the notion of 'celebration' in Ecoscenography does not only apply to those projects that have an explicit environmental message. The performance season provides an opportunity to bring visibility to the ecological values and practices of the production, no matter what genre or subject matter. As Sandra Goldmark explains, theatre designs 'tell us not only *where* we are, but *who* we are' (2017, 22, my emphasis). Our work showcases what we stand for—our material as well as our dramaturgical choices. FanSHEN's production of *Cheese* (London, 2013)—an absurdist allegory about the global financial crisis—was a wonderful example of how a bicycled-powered theatre production could still showcase sustainability without the play having an explicit ecological theme.[5] The team collaborated with the pedal power group Magnificent Revolution and local gyms to use modified exercise bikes to power their show, with audience members given discounted theatre tickets in return for their energy contribution. Every evening, the stage would plunge into darkness as the power would run out mid-show (due to pedalling power being exhausted for that day) and the switchover to the back-up electricity system would become an unexpected and crucial part of the storytelling. Building the power breakdown into the production not only was exhilarating (because no one, including the actors, knew exactly when it would happen) but also built awareness of energy expenditure in a joyful, humorous and engaging way.

Importantly, 'celebration' in Ecoscenography also applies to shows that have solemn, disturbing or confronting themes. Here, we celebrate the stage as a platform to explore those themes collectively. Consider *Sustain*, premiered at the Bergen International Festival, in Norway in 2017. *Sustain*[6] was a musical experience focused on 'sustainability, overconsumption, and the way humans have distanced ourselves from nature' through the unforeseen effects of plastic waste and consumer society (Kise 2021). The project—developed by composers Bodil Rørtveit and Jørn Lavoll, with director Vibeke Flesland Havre and designer Silje Sandodden Kise,[7] and their creative team—aimed to awaken emotions and facilitate critical reflection beyond the politics of environmental debates. Kise created a stunning immersive scenographic installation with hundreds of reclaimed plastic waste materials suspended from the ceiling and surrounding the audience (Fig. 4.6). Musicians played on specially designed plastic garbage instruments, creating a powerful aural-visual experience that was at the same time hauntingly beautiful and disturbing (Fig. 4.7). For Kise, talking to people about the work and making sure that the performance was both critical and inspiring was the biggest challenge:

Fig. 4.6 Dress of plastic garbage bags, from *Sustain* (by Bodil Rørtveit, Jørn Lavoll, Vibeke Havre, Silje Kise.) Costume design: Silje S. Kise. In picture: Bodil Rørtveit. (Photo: Jørn Lavoll)

> Many people thought that it was kind of embarrassing to make a theatre production for adults that fore-fronted environmental issues. They assumed that sustainability would not make for good art…But a few months before the premiere at Bergen International Festival in 2017, a large dead whale washed up outside our city with plastic in its stomach. This caused a big stir in the national media, and suddenly everyone was focusing on plastic and marine pollution! We had been exploring plastic waste for 3 years through our work on *Sustain*, and the show provided a timely platform for audiences to relate to the issue. (Ibid.)

Sustain demonstrated the leadership potential of theatre in providing a vital space for reflection and discussion on ecological issues. Rørtveit, Lavoll, Havre and Kise and their team created an 'empathic encounter' that activated togetherness on environmental issues, while simultaneously

Fig. 4.7 *Sustain* (by Bodil Rørtveit, Jørn Lavoll, Vibeke Havre, Silje Kise, Bergen International Festival 2017) Light design: Silje Grimstad. (Photo: Thor Brødreskift. In picture: Bodil Rørtveit)

demonstrating the resourcefulness and eco-creativity of a production with reclaimed materials.

While the artists listed above have taken highly visible and innovative ways of fore-fronting sustainability, 'celebration' in Ecoscenography can include a more subtle approach. It might entail hosting complimentary discussions, events, displays and programme notes that vanguard a production's sustainability ethic, or using social media to feature the ecological work of artists and technicians behind the scenes. No matter what the form, showcasing Ecoscenography as a valued component of our process and aesthetics—one that is fun, evocative, thought-provoking, beautiful and inspiring—not only will be instrumental to generating a greater awareness of the opportunities of sustainability, but is also at the heart of highlighting theatre's role and potential on the environmental agenda.

Act 3: Circulation

Circulation is a crucial part of ecosystem regeneration. Seed spreading, for example, is the final stage of the plant life-cycle and relies on other species and resources such as insects, birds, animals, wind and water to generate new life. For ecosystems to thrive they must be in a form of balance, which means that any resources that are consumed are either replaced or converted into other forms that are crucial to the system. Just as nature embraces waste as a valuable resource for further transformation, in Ecoscenography theatre materials and ideas are not mindlessly discarded, but considered as potential resources for sharing and co-creation beyond the production. The theatre show is the 'blooming' from which new opportunities of circulation for further making and co-creation can arise.

A key challenge of Ecoscenography is to bring theatre's resources into a cyclic model of production, where material and construction choices are based on not only creating a world in which the narrative unfolds, but also how they might be reused or distributed after the production. Thus, the role of the Ecoscenographer is to actively critique their material choices, exploring new ways of co-creating with matter in a way that speaks to this entanglement across abiotic and biotic forms. As Heim (2013) queries:

> To talk about matter is to talk about consequences, transformations and relations over time.... what is the balance between the agency of matter—as waste—and the intentions, force and design of the artist?

As Ecoscenographers we must engage with the unsettling reality that our arts practices have consequences. Nothing is 'finished' in a world of materials and there is no such thing as 'away'. Instead, skips overflow into leaching landfills and, inextricably, into our fragile ecosystems. No matter what forms they originally cast, materials are always on their way towards becoming something else (Barad 2003, 821). Where materials go after a production is a significant component of a production's environmental impact. Questioning 'what happens next' prompts new ways of approaching construction and consumption that takes the 'afterlife' of theatre materials into consideration. Ecoscenographers must rethink the potential of refuse, embracing it as a valuable resource.

'Circulation' is the least investigated area of performance design but is equally crucial to contributing knowledge, ideas, materials and experiences that can have a positive legacy beyond the theatre production. Designing with circulation in mind allows theatre makers to consider how set and costumes may be dismantled and shared for reuse. Circulation becomes a form of contribution, which involves seeking out opportunities—ways of sharing, collaborating and engaging with the world's potential (Hes and du Plessis 2014, 213). Finding ways of circulating materials to other arts organisations, recycle shops or schools allows theatre materials to be reimagined for further making, taking on new art forms and experiences that can be shared with the wider community.

Fig. 4.8 Marie-Renée Bourget Harvey's reclaimed and recirculated set design for *La Forêt, Où tu vas quand tu dors en marchant?* (Carrefour international de théâtre 2013–2014). Pictured: Joëlle Bourdon. Light: Laurent Routhier. Costume: Sébastien Dionne. Makeup: Élène Pearson & Nathalie Simard. (Photo: Francis Gagnon. After the show, everything was considered for circulation. For example, wigs were given to local Drag Queens for their own performances, and bio-materials were shredded to create compost)

The circulation of resources is a significant part of Marie-Renée Bourget Harvey's ecological stage design work. Bourget forges relationships with suppliers to purchase or borrow materials that she can then either return after the show or pass onto others, always being mindful of designing and building in such a way that allows for easy disassembly and circulation (Fig. 4.8 and Fig. 4.9). As Bourget Harvey (2021) explains:

> I am also really animated by the idea of making matter circulate, to allow it to become something else, to have several lives, several possibilities. This is something I take into account from the very beginning of the creative process, thinking about other ephemeral uses, but more often than not, I think about sustainable alternatives to facilitate re-employment. The truth

Fig. 4.9 Marie-Renée Bourget Harvey's reclaimed and recirculated design for *Madame Butterfly*, Opéra de Québec. Director: Jacques Leblanc, 2013. Light: Serge Gingras. (Photo: Louise Leblanc. The set design was made from reclaimed timber which was then returned to the supplier after the show)

> is, I'm a matter lover. I find it touching to see it circulate and meet several audiences and participate in bringing meaning from all angles.

Likewise, for stage designer and sustainability advocate Imogen Ross, circulation is a source of pride in her work. Ross has donated rugs and cushions to community groups, milk crates to a struggling social-enterprise night market and solar-powered lighting to a sustainable funeral event company as well as entire sets and costumes to small touring theatre companies. Ross' reputation for recycling and upcycling now sees local theatre makers contact her to offer set and costume materials or seek out advice on where to re-home items. As Ross (2021) highlights:

> As a young designer I was always concerned as to where the set would go at the end of the show…The emphasis was always on re-use and upcycling because it felt like we were pouring our hard-earned money into the dirt when we took things to the tip…Some designers walk away at opening night and never look back. I have never assumed it was someone else's

responsibility to solve the problems created by my sets and I always try to present upcycling or recycling pathways for my design choices…I like to know the carbon impact of my design decisions and be judged by them.

The process of thinking about what happens to their work after the production is a key part of Bourget's and Ross' creative process and artistic signature. Their approach is a demonstration of Alastair Fuad-Luke's argument, espoused in *Design Activism: Beautiful Strangeness for a Sustainable World* (2009, 189), that ecological thinking requires that we rebalance our 'focus between "what next" and "what now"'.

Avoiding the purchase of new materials by borrowing and returning resources was key to scenographer Janne Robberstad's design of Bømlo Teater's *The Salmon Surveyor* (2015). The set design featured 950 Styrofoam fish-crates, which were sourced from a local factory and provided a surface for projection as well as additional props and furniture pieces[8] (Fig. 4.10). While Styrofoam (made of plastic polystyrene) is certainly not a sustainable product, Robberstad considered the material within a closed loop cycle, where the production essentially 'borrowed' the materials to help support the telling of a local story before the Styrofoam was sent back to the nearby salmon-factory to be used directly in their manufacturing process. Robberstad's design demonstrates ways in which a non-sustainable material can be circulated through a theatre system and back into the wider world without consuming more resources. Sharing materials (rather than buying) is a simple yet effective sustainability strategy that is compatible with theatre's short timeframes.

Indeed, 'circulation' can be an opportunity for theatre artists and organisations to extend their own work, exploring opportunities (intellectual as well as material) for continuing the creative process beyond the performance event. For example, the distribution of plants for conservation was a sustainability strategy of Moira Finucane's 'climate cabaret', *The Rapture Chapter II: Art vs Extinction*[9] (Melbourne, 2019). Working with tree-planting organisation 15 Trees, a tree was planted for every table sold in the 2019 season—a total of 880 trees. The initiative was 'celebrated' and communicated to the audience during the performance season. Likewise, conservation was the focus of the set design for *Bloody, Bloody Andrew Jackson* (Indiana University, 2016), where plywood crates

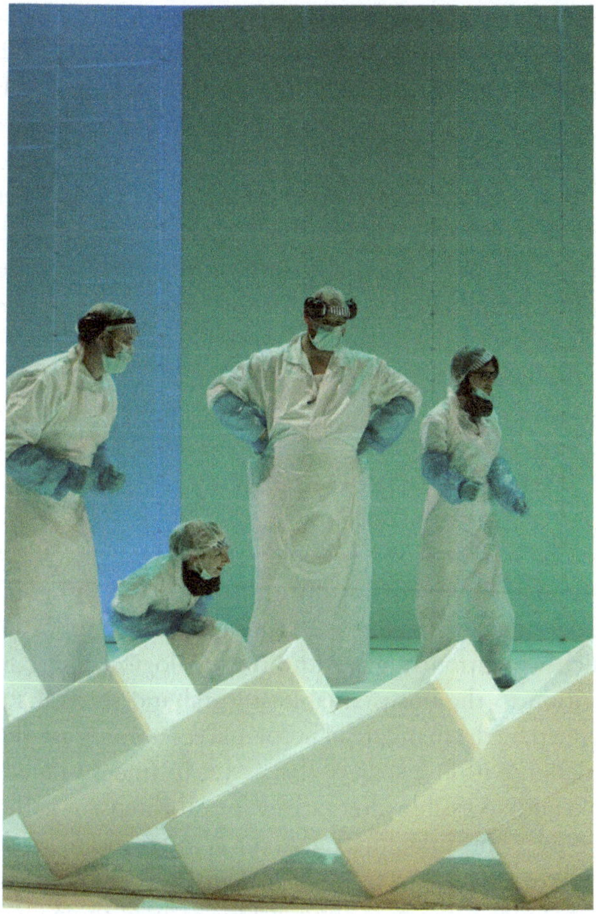

Fig. 4.10 Janne Robberstad's design of Bømlo Teater's *The Salmon Surveyor* (2015). (Photo: Janne Robberstad)

that formed part of the stage rostrum were transformed into animal homes for the University's migrating bat project (Brunner and Ranseen 2017, 25).

'Circulation' in Ecoscenography is not only about distributing theatre objects back into the system; it also includes the transferal, dissemination and contribution of ecological ideas, knowledge, partnerships and skills. Ideas of circulation and contribution were the focus of *Concierto Para El*

Bioceno (Liceu Opera, 2020, Barcelona, Spain), which took place during the global pandemic and was streamed live to the world.[10] Filling the 2292 seats of the Gran Teatre del Liceu with plants, the UceLi Quartet serenaded the leafy audience to Giacomo Puccini's *Crisantemi* to a stunning effect that was documented and circulated across the world. Sourced from local nurseries, the plants were given a short musical performance before being donated to healthcare professionals at the frontline of the COVID crisis. The project was a 'breath of fresh air' (both literally and figuratively) to people all over the world, many of whom were facing months of lockdown, and became a symbolic act of the importance of art, music and nature for health and wellbeing.

The Ecoscenography Trajectory

Ecoscenography is intended as an explorative and progressive process that involves a fundamental shift in moving from linear to cyclic modes of theatre production. This transition has significant implications for the way in which we currently make theatre. Ecological processes stand in stark opposition to the unsustainable cultural and societal structures of today, generated by the mandate of perpetual economic growth. Practising ecological design within a society that is yet to find sustainable, affordable, reliable and accessible alternatives for oil-intensive gaffa tape, synthetic fabric dyes and toxic fire retardant is a significant challenge. The transition will require a slow but radical transformation of existing structures. This includes navigating the tensions of how these changes interrelate (and undoubtedly conflict) with normative practices.

The concept of 'transitions' in sustainable development outlines how effective change-making needs to be 'radical' but not rushed (Geels and Schot 2010; Grin et al. 2010). Changes to dominant systems involve significantly reconsidering 'ways of thinking (cultures), ways of organising (structures) and ways of doing (practices)' over a period of decades (Gorissen et al. 2018, 172). Transitions occur when the 'old' system becomes more and more problematic, allowing space for the 'new' system to evolve alongside conventional practices. Changes tend to unfold across a diversity of timelines and levels, depending on existing and emerging contexts, parameters and attitudes.

Reconciling two radically different worldviews—mechanistic versus ecological—is an integral part of Ecoscenography. Theatre production requires a holistic and integrative trajectory that honours the co-evolutionary transition of moving from degenerative to regenerative modes of production. In Chap. 3, I introduced Bill Reed's (2007) Trajectory for the Whole of Sustainability as one which provides an integrated ecological approach to reconciling multiple sustainability viewpoints. Reed's proposal supports the idea of a gradual transitioning in which the designer's engagement with sustainability is explorative and progressive, trialling opportunities and engaging in acts of learning through doing. In developing a trajectory for Ecoscenography (inspired by Reed's model), my intention is to provide an aspirational path towards regenerative development, while also appreciating the co-evolution of practices that is part of this transition.

A key consideration of Reed's framework is the notion of moving from a mechanistic to an ecological worldview (Chap. 3). In the Ecoscenographic trajectory, degenerative practice (uncritical resource use, business as usual) is situated to the far left of the horizontal scale, with regenerative practice situated at the peak of the trajectory. Here, the concept of 'regenerative' not only implies 'reinvigorating or reviving a system', but also 'changing the system into something different and better' that also informs a 'new way of being' (Hes and du Plessis 2014, 111). Regenerative practice implies an 'active, contributive engagement with the world' where the designer's focus is on creating a positive ecological footprint that contributes to environment and community (Hes and du Plessis 2014, 10). This is the transition to what McDonough and Braungart (2002) describe as moving beyond 'doing less bad' to 'doing more good'. As the shift in ecological *thinking* deepens, so does the potential for ecological *practice*. The iterative cycle of 'co-creation—celebration—circulation' helps increase ecological knowledge, awareness and engagement, which strengthens with each sequence of practice.

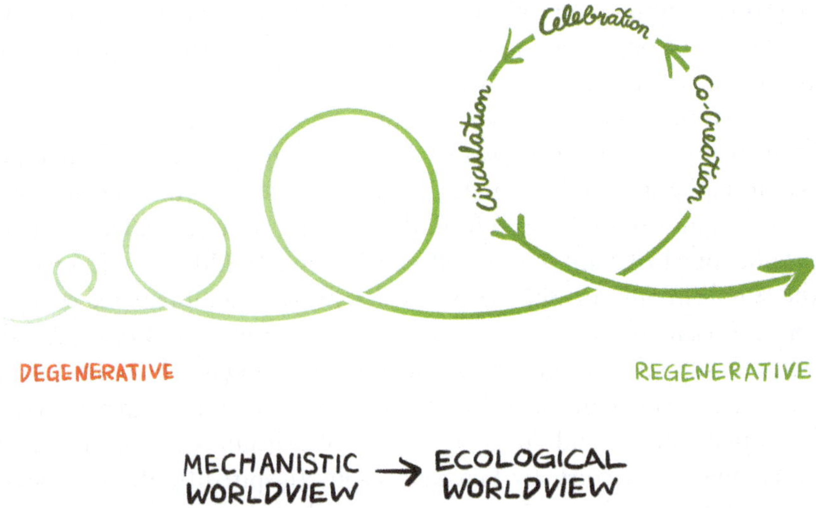

Ecoscenography is intended as an evolving, experimental, progressive and transversal process which considers how things, beings, places and systems evolve and co-emerge in relationship to one another. Rather than a strict progression or linear pathway (as represented here for clarity), the framework is one of cycling back and forth as part of transitioning from a mechanistic to an ecological worldview. Envisioned as inclusive, flexible and expansive, Ecoscenography starts with setting the intention to work in an ecological way, no matter where one is situated across the path. Essentially, the Ecoscenography trajectory is about transforming ecological values into action, where the shift to ecological practices is one of continuous unfolding and opening up of eco-creative consciousness. It is thinking globally but starting locally. It is thinking big but starting small. As scholar, quantum physicist and environmental activist Vandana Shiva highlights:

> You begin with the small, you begin with the do-able, you begin with that which you can do and that which you can do under the most repressive context. And in the doing of it, you make the repression retreat, you change the terms…To make it happen takes personal commitment. It has to begin with the person, it has to begin at the local. … what you bring to life—

whether you do it as an artist or you do it as conservationist—is the total function of how much you believe it can be done. (2014, 27, 29)

Challenges and Opportunities of Ecoscenography

While Ecoscenography aims to offer a holistic approach to bringing sustainability into theatre production, its success will undoubtedly rely on how much the team of designers, producers, directors, stage managers, artistic and technical directors give value to its potential. As discussed in Chap. 2, the limitations of ecological design are largely determined by the paradigm, context and environments in which theatre artists are working. Performance making is a highly collaborative art form, and as such, there are a number of personal and organisational factors that can make ecological practice difficult. This is further exacerbated by the fact that most organisations in the performing arts do not yet have sustainability policies in place, making it harder for teams to implement these approaches. These contexts will require a stronger level of determination and commitment than others. As Goldmark (2017, 22) writes, 'our patterns of consumption are a choice. But, like many of our patterns of behaviour, choice is limited by many factors, including speed, budget, and, perhaps most of all, habit'. In many cases, designers who find themselves in the hands of a dedicated director or production manager will have a better chance at implementing Ecoscenography than those where sustainability is not a priority.

Nevertheless, there are obvious challenges for pursuing an ecological worldview in a mechanistic paradigm. As discussed above, conventional theatre paradigms often contradict more holistic and organic ways of working. Hard deadlines and limited budgets frequently inhibit abilities to question and test new ways of doing things. In addition, confronting unsustainable practices requires confidence, determination and commitment. A collaborative dedication to ecological values is necessary to create a more conducive environment for change.

Time is a major issue for pursuing ecological choices, particularly in production week—a critical time when all the elements (set, sound, light, action) come together to create the 'magic' of the show. This is often a messy and chaotic time where the looming deadline of Opening Night is crucial and the mandate is 'just get the job done'. It is at this moment when sustainable considerations are likely to be undermined—when designers and technicians resort to using toxic products, such as fast drying spray paint to deal with aesthetic problems (often in the hazardous environment of a non-ventilated setting). The key is to be prepared for these potential challenges. As Ecoscenographer Andrea Carr (2015) contends:

> There is always a balancing act to consider in any of these ventures—the use of fuel, water and electricity as well as extra support and time. Eco-driven initiatives can take longer than going out and making a direct purchase or even making something from scratch. A deeper understanding and reframing of design and production practices is of vital importance. The materials may cost less… but the labour can be more intense. I believe this needs to be reflected in how budgets are calculated and people's skills and time valued.

Despite its potential, bringing an ecological approach into theatre production entails questioning unsustainable practices, perceptions and assumptions as part of the creative process. In reality, this means understanding that each project comes with its own collaborators, agendas, parameters, policies and processes which are often varying and contradictory (Robertson 2014, 314). There are no easy recipes or ten-step plans (Hes and du Plessis 2014, 212). Ecological practice requires the ability 'to critically reflect on your own thinking, being constantly vigilant against logical fallacies, dangerous assumptions and specious argumentation' (Ibid.). Even within more conventional practices, challenging preconceived ideas—such as questioning the assumption that certain materials, products and services are more sustainable than others—is an important start to engaging in Ecoscenography. Sustainability requires the ability to question existing structures and paradigms while embracing the autonomy of finding one's own path through the complexities.

How far theatre makers are willing to go in the pursuit of ecological practice is highly personal and largely dependent upon values, privilege, circumstances and opportunity. In an industry where sustainability policies are still a rarity, ecological design is largely seen as a personal investment which only exacerbates the heavy workload that designers and other collaborators regularly maintain and for which they are rarely compensated. Bringing ecological practice into conventional practice will continue to be problematic as long as organisations, colleagues and educators do not see it as a value. Campaigning for structural changes that support ecological practice and taking pride in the satisfaction of any sustainability achievement (big or small) are essential. As set designer Justin A. Miller posits, every production is 'an opportunity to be more environmentally conscious' which 'involves thinking ahead, remembering the big picture, and resisting the "way-it's-always-been-done" in favor of innovation'—a process which requires as much creativity as it does organisation (Miller 2012, 199).

No matter what the context, I contend that there are ways for designers to engage with ecological ideas that open up wider perspectives and opportunities. My conceptualisation of Ecoscenography is intended as one that is open to a diversity of theatre-making approaches and aesthetics across cultures, with a framework that does not dictate rules, checklists or specifications of practice, but rather nurtures serendipitous exploration, experimentation and collaboration. Importantly, Ecoscenography is not about 'getting it right' the first time or even the 20 times after that. It is not about reprimanding others or wallowing in the state of our unsustainable affairs. Ecoscenography simply starts with the conscious and active choice of wanting to make work that is in tune with our socio-ecological values and a willingness to start that journey, despite the challenges ahead. As stage designer and sustainability advocate Soutra Gilmour explains, 'You're always imperfect—that is for sure. It's a constant picking oneself up on the things you are and aren't doing—and celebrating the successes' (Gilmour in Julie's Bicycle 2013, 6)

Whilst I am excited about the possibilities, I am also aware of the binaries and limitations that this concept may create. The very separation of Ecoscenography as something 'other' from scenography may generate a perception of exclusivity which only reaffirms the binaries that this book is trying to disrupt. Further, distinguishing 'Ecoscenography' from scenography may imply that work done outside of 'Ecoscenography' is not ecological, which is undoubtedly not the case.

Nevertheless, I propose that if the performing arts were already transitioning to an ecological model, then the need to define Ecoscenography as a separate practice would not be necessary. Yet theatre production is not yet at a point where 'sustainable design' is synonymous with 'good design'. Thus, I maintain that the distinction between Ecoscenography and scenography is useful, as it draws attention to the 'eco' and frames its importance. As to the future of the term, I welcome the possibility that 'Ecoscenography' will one day be superfluous and therefore no longer required.

While Ecoscenography may seem like a daunting concept, this is also where innovation and possibility lie. Adopting an ecological approach allows theatre makers to imagine ideas well beyond the performance—to consider how ideas of 'creative expansion' (a widening of identity in ecological thinking) can make a contribution to the broader world. Ecoscenography works from the premise that there is nothing more fulfilling than connecting to the living world in a way that is fundamental, positive and inspiring. The following chapter provides insights into what Ecoscenography might look like if we approach sustainability from a place of abundance, rather than one of scarcity and limitations. It demonstrates opportunities for integrating creativity, aesthetics and innovation with sustainable practice. Using examples of my own work, I explore how Ecoscenography can be a mode of more-than-human engagement that can create ripples into wider systems, where performance design can act as a force for regeneration.

Notes

1. For more information about Ruth Stringer's work, see: https://www.ruthstringer.com.
2. For more information about Donyale Werle's work, see: http://www.donyalewerle.com.

3. For more information about Andrea Carr's work, see: http://www.andreacarr.co.uk.
4. For more information about *Tejidos* by Aris Pretelin-Estéves, see: https://sitioespecifico.wixsite.com/tejidos/pieza-escenografica.
5. For more information about FanSHEN's production of *Cheese*, see: www.fanshen.org.uk/cheese. FanSHEN recently changed their name to Fast Familiar. Their latest work can be viewed via: https://fastfamiliar.com.
6. For more information about *Sustain*, see: www.sustaintheconcert.com.
7. For more information about Silje Sandodden Kise's work, see: www.siljekise.no.
8. For more information about the *Salmon Surveyor*, see https://ecoscenography.com/2016/03/21/the-salmon-surveyor-by-janne-robberstad. For more information on Janne Robberstad's work, see: https://www.spindelmaker.com.
9. For more information about Moira Finucane's 'climate cabaret', *The Rapture Chapter II: Art vs Extinction* (Melbourne, 2019), see: https://www.finucaneandsmith.com/the-rapture.
10. For more information about *Concierto Para El Bioceno* (Liceu Opera, 2020, Barcelona, Spain) see: https://www.liceubarcelona.cat/en/concierto_bioceno.

References

Avidar, Noémie. 2020. Place-based Ecoscenography: Interview with Noémie Avidar. In *Ecoscenography: Adventures in a New Paradigm for Performance Making*, ed. Tanja Beer. https://ecoscenography.com/2020/12/01/place-based-ecoscenography-interview-with-noemie-avidar/.

Bal, Mieke, and Christine van Boheemen. 2009. *Narratology: Introduction to the Theory of Narrative*. Toronto, ON: University of Toronto Press.

Banks, Summer. 2014. Riding a Bike on Stage: Green Theater and Katie Mitchell's Production of Lungs. In *Howlround Theatre Commons*. https://howlround.com/riding-bike-stage.

Barad, Karen. 2003. Posthumanist Performativity: Toward an Understanding of How Matter Comes to Matter. *Gender and Science: New Issues* 28 (3): 801–831.
Benyus, Janine. 2002. *Biomimicry: Innovation Inspired by Nature*. New York: HarperCollins.
Bourget Harvey, Marie-Renée. 2021. Joyful Eco-creativity: Interview with Marie-Renée Bourget Harvey (Canada). In *Ecoscenography: Adventures in a New Paradigm for Performance Making*, ed. Tanja Beer. https://ecoscenography.com/2021/01/28/joyful-eco-creativity-interview-with-marie-renee-bourget-harvey-canada/.
Brunner, Paul, and Olivia Ranseen. 2017. The Greening of Academic Theatre: While Several Impediments Hinder Successful Long-Term Integration of Sustainable Practices, They Aren't Insurmountable. *TD&T (Theatre Design & Technology)* 53 (3): 20–33.
Carr, Andrea. 2015. Sleeping Bag Metamorphosis. In *Ecoscenography: Adventures in a New Paradigm for Performance Making*, ed. Tanja Beer. https://ecoscenography.com/2015/12/01/sleeping-bag-metamorphosis-by-andrea-carr/.
Fair Share Fare. 2019. Portage. In *Fair Share Fare*. https://www.fairsharefare.com/portage.
Fried, Larry, and Theresa May. 1992. *Greening Up Our Houses: A Guidebook to an Ecologically Sensitive Theatre Organization*. New York: Drama Book Publishers.
Fuad-Luke, Alastair. 2009. *Design Activism: Beautiful Strangeness for a Sustainable World*. London, UK: Earthscan Publications Limited.
Geels, Frank, and Johan Schot. 2010. The Dynamics of Transitions: A Socio-Technical Perspective. In *Transitions to Sustainable Development: New Directions in the Study of Long Term Transformative Change*, ed. John Grin, Jan Rotmans, and Johan Schot, 9–87. New York, USA: Routledge.
Goldmark, Sandra. 2017. Theatre 'Stuff:' Use and Reuse in the Theatre: Examining our Human Entanglement with Stuff and Mitigating its Environmental Impact. *TD&T (Theatre Design & Technology)* 53 (1): 20–29.
Gorissen, Leen, Felix Spira, Erika Meynaerts, Pieter Valkering, and Niki Frantzeskaki. 2018. Moving Towards Systemic Change? Investigating Acceleration Dynamics of Urban Sustainability Transitions in the Belgian City of Genk. *Journal of Cleaner Production* 173: 171–185.
Grin, John, Jan Rotmans, and Johan Schot. 2010. *Transitions to Sustainable Development: New Directions in the Study of Long Term Transformative Change*. New York: Routledge.
Hann, Rachel. 2018. *Beyond Scenography*. London, UK: Routledge.

Haraway, Donna. 2016. *Staying with the Trouble*. Durham, NC: Duke University Press.

Heim, Wallace. 2013. Investigating the Aesthetics of Ecological Design and Eco-Scenography. In *World Stage Design: People, Planet and Profit*. Cardiff, Wales. http://wallaceheim.com/investigating-the-aesthetics-of-ecological-design-and-eco-scenography/.

Hes, Dominique, and Chrisna du Plessis. 2014. *Designing for Hope: Pathways to Regenerative Sustainability*. Hoboken, NJ: Taylor and Francis.

Julie's Bicycle. 2013. Sustainable Production Guide. Accessed November 24, 2021. https://juliesbicycle.com/resource-production-guide-2013/.

Kimmerer, Robin Wall. 2013. *Braiding Sweetgrass: Indigenous Wisdom, Scientific Knowledge, and the Teachings of Plants*. Minneapolis, MN: Milkweed Editions.

Kise, Silje Sandodden. 2021. Performing Sustainability: Interview with designer Silje Sandodden Kise (Norway). In *Ecoscenography: Adventures in a New Paradigm for Performance Making*, ed. Tanja Beer. https://ecoscenography.com/2021/01/18/performing-sustainability-interview-with-designer-silje-sandodden-kise-norway/.

Kompanek, Christopher. 2012. For Donyale Werle, It's Easy Being Green. In *American Theatre: A Publication of Theatre Communications Group*. https://www.americantheatre.org/2012/09/01/for-donyale-werle-its-easy-being-green/.

Mang, Pamela, and Bill Reed. 2012. Designing from Place: A Regenerative Framework and Methodology. *Building Research & Information* 40 (1): 23–38.

May, Theresa. 2013. *Salmon is Everything: Community-Based Theatre in the Klamath Watershed*. Corvallis, OR: Oregon State University Press.

McDonough, William, and Michael Braungart. 2002. *Cradle to Cradle: Remaking the Way We Make Things*. New York: North Point Press.

McKinney, Joslin. 2015. Scenographic Materialism, Affordance and Extended Cognition in Kris Verdonck's ACTOR #1. *Theatre and Performance Design* 1 (1–2): 79–93.

McKinney, Joslin, and Scott Palmer. 2017. *Scenography Expanded: An Introduction to Contemporary Performance Design*. London, UK: Bloomsbury Methuen Drama.

Miller, Justin. 2012. The Labor of Greening Love's Labour's Lost. In *Readings in Performance and Ecology*, ed. Wendy Arons and Theresa May, 191–201. New York: Palgrave Macmillan.

Nelson Education Ltd. 2004. Aboriginal Perspectives. In *Nelson Education Ltd*. https://www.nelson.com/aboriginalstudies/aboriginalperspectives/about.html.

Pascoe, Bruce. 2014. *Dark Emu, Black Seeds: Agriculture or Accident?* Broome, Australia: Magabala Books.
Pearson, Mike. 2018. Site-Specific Theatre. In *The Routledge Companion to Scenography*, ed. Arnold Aronson, 295–301. New York: Routledge.
Pedersen Zari, Maibritt. 2016. Mimicking Ecosystems for Bio-Inspired Regenerative Built Environments. *Journal of Intelligent Buildings International* 8 (2): 57–77.
Pretelin-Esteves, Aris. 2021. Encounter-spaces and the City as Scenography: Interview with Aris Pretelin-Esteves (Mexico). In *Ecoscenography: Adventures in a New Paradigm for Performance Making*, ed. Tanja Beer. https://ecoscenography.com/2021/02/03/encounter-spaces-and-the-city-as-scenography-interview-with-aris-pretelin-esteves-mexico/.
Rawes, Peg. 2013. *Relational Architectural Ecologies*. Oxfordshire, UK: Taylor & Francis Ltd.
Reed, Bill. 2007. Forum Shifting from 'Sustainability' to Regeneration. *Building Research & Information* 35 (6): 674–680.
Robertson, Margaret. 2014. *Sustainability: Principles and Practice*. Vol. Book, Whole. London: Routledge/Taylor & Francis Group.
Ross, Imogen. 2021. Resourceful Ingenuity: Interview with Performance Designer Imogen Ross (Australia). In *Ecoscenography: Adventures in a New Paradigm for Performance Making*, ed. Tanja Beer. https://ecoscenography.com/2021/01/31/resourceful-ingenuity-interview-with-performance-designer-imogen-ross-australia/.
Shiva, Vandana. 2014. Beginning with the Small. In *Landing Stages: Selections from the Ashden Directory of Environment and Performance 2000–2014*, ed. Wallace Heim and Eleanor Margolies, 24–29. London, UK: Crinkle Crankle Press.
Stringer, Ruth. 2021. Earth-based Serendipitous Scenography: Interview with Set and Costume Designer Ruth Stringer (UK). In *Ecoscenographer: Adventures in a New Paradigm for Performance Making*, ed. Tanja Beer. https://ecoscenography.com/2021/01/21/earth-based-serendipitous-scenography-interview-with-set-and-costume-designer-ruth-stringer-uk/.
Van der Ryn, Sim, and Stuart Cowan. 2007. *Ecological Design: Tenth Anniversary Edition*. 10th Anniversary edn. Aufl. Washington, DC: Island Press.

5

Two Case Studies in Ecoscenography

T. Beer, *Ecoscenography*, https://doi.org/10.1007/978-981-16-7178-4_5

As an Ecoscenographer, it is my desire to deepen connections with the more-than-human world that has greatly influenced the way I practise. I seek to create performance spaces that enrich not only audiences, but also the communities and environments in which they are situated. Much of my recent practice has taken place in the realm of expanded scenography, where opportunities to push boundaries have proven greater than within the confines of conventional theatre. Furthermore, in stepping outside of traditional theatre venues and processes, I have been free to explore new practices and aesthetics that support performance designers as contributors and change agents. This has been a necessary choice for my conceptualisation of Ecoscenography, as one which spans both traditional and expanded practice, and opens up new ways of thinking and doing.

A major incentive of my work has been to push against long-held assumptions of sustainability in the performing arts: the notion (as I outline in the *Introduction*) that ecological design is expensive, boring, time-consuming and limiting of high-quality aesthetics. I am driven to demonstrate the potential of Ecoscenography—one where sustainability is celebrated, not a 'conversation stopper' or a hidden practice that is kept behind closed doors. To do this, I have largely found myself operating outside my traditional role as a performance designer to take on new responsibilities—as a theatre maker, producer and community facilitator. My expanded practice has led to expanded roles. Roles that I never intended to undertake when I first trained as a set and costume designer, but nevertheless have become a necessary part of my desire to achieve the regenerative practices.

This chapter presents two case studies in Ecoscenography, *This Is Not Rubbish* and *The Living Stage*. Each sought to foster the cycle of co-creation—celebration—circulation to generate a 'gesamtkunstwerk' or total artwork that embraces and celebrates the making and unmaking of theatre production within the realm of expanded scenography. At the heart of these projects is the notion of incorporating regenerative strategies—community-engaged and contributive approaches—that reimagine theatre production for an ecological paradigm. Guided by Integral theorist Ben Haggard's (2002, 31) concept of 'ongoing potential', the projects act as continually unfolding sources of inspiration and spirit that uncover possibilities for practising Ecoscenography. While these projects were conducted in the realm of the expanded, my hope is that they hold valuable insights into eco-creative processes that can also be transferred to more conventional theatre contexts.

As discussed in my chapter outline of the book, it is my intention that readers will first view the supporting film recordings and additional photos that are available via the Ecoscenography website (www.ecoscenography.com) before reading this chapter. This will not only provide readers with a stronger introduction to the scenographic experience (the materials and collaborators involved, the performances and the aesthetic outcome) but also provide a space for multiple viewpoints. For *The Living Stage* in particular, this is to give agency to, and show respect for, the many voices that have been instrumental to the success of these projects. While this decision might be seen as unorthodox, I do this to acknowledge the limitations of my own biases, descriptions and analyses of the work and provide a broader spectrum for the reader to understand it. The full list of collaborators is provided in the project credits at the end of this book.

Assuming that the reader has become familiar with the works through the supporting online documentation, my intention for this chapter is not as a step-by-step detailed description of the projects, but to describe how they fit with the conceptualisation of Ecoscenography outlined in the preceding chapters.

This Is Not Rubbish

This Is Not Rubbish (Melbourne, Cardiff, London, 2012–2013) was a self-initiated project that explored the journey of a material rescued from landfill, and its capacity to create immersive performance spaces and wearable artefacts. The aim of the work was to 'challenge preconceptions about what is considered rubbish, as well as the need for scenographers and performance makers to continuously begin anew in a world of increasing environmental concerns' (Beer 2016b, 493). First commissioned by the Arts Centre's Riverside Live programme (Melbourne), the work included three devised performance installations (entitled *Strung*) which were realised at the Hamer Hall (Melbourne), World Stage Design

(Cardiff) and the Royal Central School of Speech and Drama (London). Part Ecoscenographic demonstration and part performance installation, *Strung* was performed with one bag of reclaimed salami netting offcuts, one performer and a team of active scenographers who worked together to build an installation directly in front of the audience. After the performance, the installation was knitted into unique fashion accessories and auctioned for charity. Central to the project was the exploration of how recycled materials and found objects could be circulated through the scenographic process—to make a contribution in and beyond the production. The focus was one of human-material co-creation where designers collaborate with abiotic substances as partners of transformation (Beer 2018).

Co-creating with Matter

My initial commission for *This Is Not Rubbish* was to create an Ecoscenographic installation that would activate the foyer of the Hamer Hall. The open brief provided an opportunity to respond directly to the structural elements of the space and to accentuate its grand architecture. Inspired by the notion of finding a string-like material that could link the columns of Hamer Hall foyer, I made a visit to Reverse Art Truck—a much-loved local recycling warehouse for artists—to activate my imagination.

Cultural theorist Liz Parsons (2008) describes the act of choosing an object as a form of creative operation. In particular, it is our serendipitous relationship with found objects that opens up opportunities for co-creation. Walking through Reverse Art Truck in pursuit of a material, I recall a rush of excitement and a feeling of creative possibility. The warehouse's array of rescued factory offcuts and unwanted oddities destined for landfill (from timber rods, bottle tops and foam blocks to spools of metallic papers, vintage wrapping, upholstery fabric and trims) opened up a spectrum of potential that allowed 'objects to speak to me through their materiality, rather than what had been imposed on to them by human semiotics and societal expectations' (Beer 2018, 111).

Anthropologist Tim Ingold (2012, 435) argues that 'anything we come across could, in principle, be regarded as either an object or a sample of material'. The act of finding required my ability to seek out the potentiality of things as both 'objects' and 'materials'. In other words, when I viewed something as an object, I saw it in terms of its past use, functionality, identity and status (Miller 1995, 41). In viewing the same thing as a sample of material, however, was to see it as an actant for further making and transformation (Ingold 2012, 435). These two distinct 'ways of seeing' allowed me to engage with a variety of responses that helped to activate my imagination. It was by opening myself up to the variability of material oddities at Reverse Art Truck that allowed me to find the salami netting[1] that I ultimately chose for the design. While the netting may not have been what I originally had in mind, exploring the material's tightly woven elasticity, softness, strength and flexibility signified an exciting avenue for scenographic exploration (Fig. 5.1).

Co-creating with matter is to recognise its role as a partner in activating Ecoscenographic potential. This is not about engaging with the hylomorphic model of making which assumes the artist as an agent imposing

Fig. 5.1 Salami netting material in its original form (*This Is Not Rubbish*, Melbourne, 2013). (Photo: Geoffrey Heard)

form on an inert or biotic substance that is incapable of having its own agency (Ingold 2010, 91). Instead, co-creation is seen as a process of growth, or ontogenesis, which requires a change of focus from the emphasis on the 'objectness of things', to an engagement with the qualities and possibilities of the material-collaborator (2012, 431). Ingold describes making as 'a process of correspondence' that requires the 'bringing forth' of a material's potentiality (2010, 97, 2012, 435). One of engaging with 'the matter-flow', of 'surrendering' to the material, to the more-than-human collaborator and 'following where it leads' (Deleuze and Guattari 2004, 408–409, 454). Here, the scenographer opens themselves up to the 'worlding' of materiality and complexity in the greater context of the more-than-human world where they are no longer a 'space manipulator' but a mediator, facilitator and collaborator (Beer 2016a, 170).

Co-creating with the netting material on *This Is Not Rubbish* became one of material-human conversation in which the first signs of the object's characteristics and potential began to be revealed. The salami netting was cut and knotted, taking on a new shape of ongoing and unexpected formation (Fig. 5.2). It was here that I brought my movements in 'close

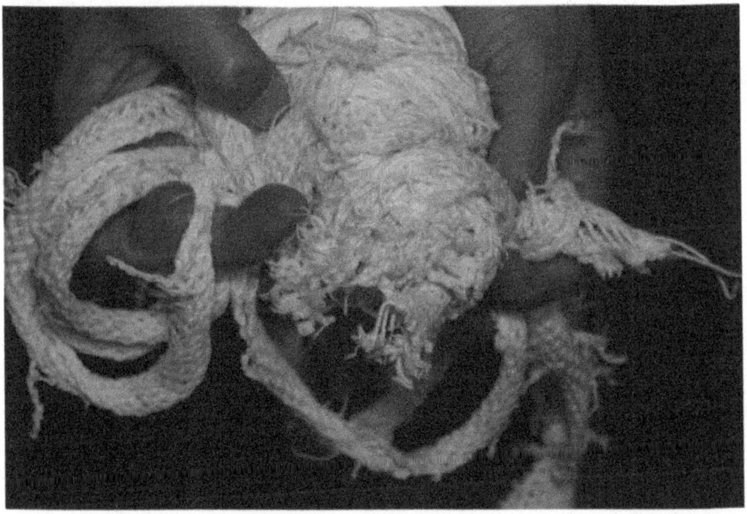

Fig. 5.2 Starting the transformation of salami netting material (*This Is Not Rubbish*, Melbourne, 2013). (Photo: Geoffrey Heard)

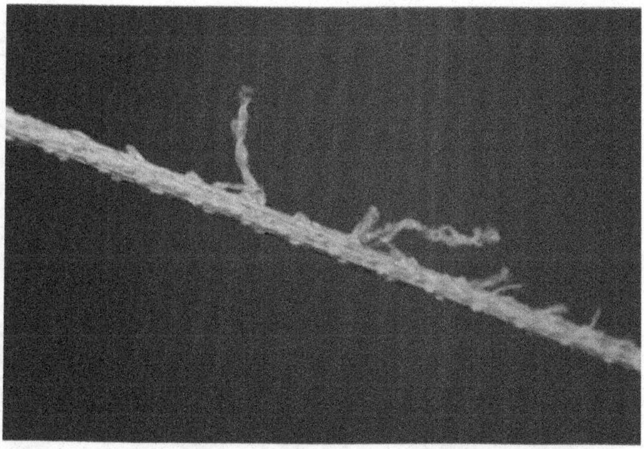

Fig. 5.3 Transformation of salami netting into a string-like material (*This Is Not Rubbish*, Melbourne, 2013). (Photo: Geoffrey Heard)

correspondence' with those of the material, becoming 'mutually entangled' in creating a web around myself and the furniture (Ingold 2012, 436, 438). By aligning myself with the netting material and surrendering to its resistances and offers, each gesture became a question to which the material responded according to its qualities. In this act of material-human partnership, I was able to transform the salami netting into a string-like material for further scenographic exploration (Fig. 5.3).

Human-material co-creation is one of collaboration and respect, breaking apart hierarchical assumptions of materials and processes, to give value to their agencies and potential as part of our creative process. As Ingold describes, 'forms of things arise within fields of force and flows of material' and it is through 'joining with and following the forces and flows' that bring a work to fruition (2010, 91, 2012, 435). However, in this act of following the 'matter-flow' (Deleuze and Guattari 2004, 371), there was also uncertainty—a concern that the overall function and aesthetic value of the material was not pleasing and a nervous anticipation of what might emerge. This feeling of 'uncertainty' and 'not-knowing' is also articulated by Donyale Werle, who describes the turbulent process of working with salvaged materials in her interview in American Theatre: 'I work on the edge of disaster a lot, and eventually it turns. It's nerve-wracking because you don't know when that'll happen' (Werle in Kompanek 2012, 30).

Fig. 5.4 Material-spatial investigations for *Strung* (*This Is Not Rubbish*, Melbourne, 2013). (Photo: Geoffrey Heard)

It was only after several hours of cutting and tying that the material for *Strung* began to unveil its hidden potential for me, taking on an unexpected quality of beauty and strength. From my first experiences of uncertainty, the netting material was transformed into scenographic potential (Fig. 5.4). It offered more scope than previously imagined (Fig. 5.4).

Celebrating Human-Material Assemblages

A key objective of *Strung* was to use 'installation-making as improvisational performance' to 'celebrate' the potential and beauty of a discarded material in the public realm. Working with the netting material as a 'more-than-human ensemble', the Melbourne team (Jennifer Tran, Ryan Foote, Sabrina D'Angelo and myself) constructed the string-like installation directly in front of the audience, transforming the space through the material-bodily assemblages created. *Strung* broke through traditional

Fig 5.5 Performing *Strung* (Hamer Hall, Arts Centre, Melbourne, 2013). (Photo: Gisela Beer)

designer and performer roles, producing a connection of simultaneous making and performing as we literally spun a web of material between us (Fig. 5.5).

After its initial performance at Hamer Hall in Melbourne, *Strung* was reimagined and recycled for two further presentations in Cardiff and London (as part of my two-year sabbatical in the UK) with each performance environment influencing the way the work was perceived. Spanning across continents and regions, each performance was adapted to the place in which it was presented, including responding to the unique set of spatial conditions provided and inviting local designers and performers to collaborate on the project. For example, the second performance took place in the exterior grounds of the Royal Welsh College in Cardiff as part of World Stage Design (WSD), where I invited WSD performers and participants to take part in the making of the installation. I incorporated trees on site as natural 'pillars' to support the design and presented the work as a scenographic demonstration rather than a performance to better suit the WSD programme.

The final performances of *Strung* took place in a black box studio space at the Royal Central School of Speech and Drama (hereafter, 'Central') in London (Fig. 5.6). Responding to place inside the theatre allowed us to make use of the existing lighting rig and stock as a way of maximising both economic and environmental efficiencies. Using what was already available at Central gave us access to many more scenographic possibilities. Extra elements such as focused lighting, black curtaining and the

Fig. 5.6 Circular set up for *Strung* (Royal Central School of Speech and Drama, London, 2013). (Photo: Alex Blake)

addition of a live musician brought a higher level of focus to the work. The overhead rig permitted us to hang the material from above the audience (not just relying on existing vertical points such as architectural elements), thus allowing for a larger and more dynamic canopy to take shape. Having multiple attachment points maximised the scope of the design to create a stunning material web.

The Central performance began with the audience seated at the circumference of a circular edifice of white string, dangling from the ceiling and surrounding vertical pillars. Guitarist Harry Webber accompanied performer Christina Kapadocha as she moved into the centre to begin her dance with the string. Dressing herself in a string costume made from six tubes of reclaimed cotton-elastic material, Kapadocha's movements, instincts and responses to the string were watched carefully. Once dressed, four 'active scenographers' (Ella Marie Fowler, Jacquie Holland, Natalie Jackson and myself) entered the space and began connecting Kapadocha to the suspended canopy, one piece of string at a time (Fig. 5.7). Working

Fig. 5.7 Performing *Strung* (Royal Central School of Speech and Drama, London, 2013). (Photo: Alex Blake)

Fig. 5.8 Spinning a web of material in *Strung* (Royal Central School of Speech and Drama, London, 2013). (Photo: Alex Blake)

as seven ensemble members (one performer, one musician, four active scenographers and one bag of string material) we allowed for human and non-human actants to inform each other in our embodied experience (Fig. 5.8). By placing ourselves as co-performers in the confines of a focused space, we were able to fully explore the merging of more-than-human actants (material, bodily and spatial) as part of the scenographic experience (Fig. 5.9). We were improvising live, simultaneously becoming scenographers and performers, 'knotting strings together and exploring the infinite possibilities of the growing web' (Beer 2016a, 166). As we worked, we allowed each string's individual length, thickness and elasticity to dictate how we moved and constructed the shape of the web. Stretching the string across the space, we were pursuing 'the "hinge" between bodily movement and abstract reason, between the textilic and the architectonic, between the haptic and the optical, between improvisation and abduction, and between becoming and being' (Ingold 2010, 100) (Fig. 5.10). Soon after 'each piece of string had been connected, the

Fig. 5.9 Leaning into the material in *Strung* (Royal Central School of Speech and Drama, London, 2013). (Photo: Alex Blake)

Fig. 5.10 Final *Strung* performance (Royal Central School of Speech and Drama, London, 2013). (Photo: Alex Blake)

Fig. 5.11 Final *Strung* installation after the performance (Royal Central School of Speech and Drama, London, 2013). (Photo: Alex Blake)

participants departed to the periphery of the space leaving Kapadocha to emerge out of the web, delicately releasing herself from the string, one costume piece at a time, like a butterfly from a cocoon' (Beer 2016a, 166). As she exited the space, 'her presence was left by the shape of the web that filled the stage, beautifully intact with elongated pieces of her costume debris left pendulous in the space' (Ibid.). The installation became a physical record of the complex agentic assemblages shared (Fig. 5.11).

Strung broke through traditional designer and performer roles, producing a connection of simultaneous making and performing as we literally spun a web of material between us. Audiences were invited into the space to join us in the making of the web and engage with the tactility of the material. Each contributor brought their own ideas into the realm offering 'new statements on the visible sculpting of space in real time' (Beer 2016b, 503). By integrating spectators into the performance experience, we were inviting the public to witness how a material might have agency and how this potential vibrancy could allow for a renewed appreciation of the material, far beyond its original purpose or landfill destination. Here, borders between performers, material and the public were made permeable, opening up more-than-human sensitivities. A clear intention of the project was to invite participants to immerse themselves in the experience of becoming more conscious of the material's liveliness—to see themselves as more-than-human assemblages (Bennett 2010). In this way, *Strung* aimed to interlink Ecoscenographic notions of co-creation with celebration.

Once the installation was complete, audiences were invited to join us in a circle underneath the weblike canopy to discuss the project, its process, aesthetics and material viewpoints. Q&A style conversations were centred on topics of sustainability and human-material relationships, and how we give value to things, while discarding others. Themes of overconsumption, plastic islands and landfill sites were offset by the absurdity of finding out that the signature material was discarded (unused) salami netting, which brought much-needed humour and creativity into the conversation. Our intention was to move past the guilt and despair, to focus on opportunities of material human collaborations. *Strung* was not only a 'celebration' of the beauty and value of discarded material but also an opportunity to advocate for more considered materialist choices that recognised the co-extensive effects of matter.

Circulating Value

In the final performances of *Strung*, the salami netting was beginning to be stretched beyond its capacity, snapping often when pulled on abruptly. The material was losing its strength and beauty, becoming more dishevelled with each performance—it was signalling that it was time to move on and explore other potentialities. Exploring the new reality of the netting, we found its former elasticities and capacities (the very characteristics that had drawn me to the co-creative possibilities of the netting in the first place) had slackened and transformed.

Ecoscenography acknowledges matter 'as an active participant in the world's becoming' (Barad 2003, 803). 'Circulation' in Ecoscenography is about recognising that material entanglement occurs across bodies, substances and environments with co-extensive effects that perform well beyond the theatre. Becoming alert to the material's co-creative capacities and sensitivities calls for a renewed appreciation for materials, where its future potential—its vibrancy and co-creative capacities—is embraced through what I regard as 'expansive listening' (Beer 2016a, 2018). 'Expansive listening' is evident in ecomaterialist ideas of perception which aim to 'consult non-humans more closely, or to listen and respond more carefully to their outbreaks, objections, testimonies, and propositions' (Bennett 2010, 108). The act of 'listening' to the netting material allowed me to consider how it might evolve next—to further 'extract' and 'release' the potential of the material and investigate its intra-active state of becoming beyond the performance season (Barad 2003, 828; Beer 2016a, 169). For Ingold (2011, 3), nothing is ever finished in a world of materials: 'everything may be something, but being something is always on the way to becoming something else'. It is through the designer's capacity to see materials not as inert, but as continuous and agentic 'avenues for becoming' that uncovers a vast array of wondrous Ecoscenographic ideas for circulation (Beer 2016b, 492). After the final showing of *Strung*, a craft circle was set up at Central to knit the material into wearable art and auction the items for charity.

Various platforms have demonstrated how the act of 'crafting together' allows groups to make meaningful connections, create collective identities, which in turn may assist in communities taking a hands-on approach to tackling ecological and social issues more collaboratively. From the *Strung* material, the craft group worked together to knit a diverse selection of art accessories. A series of accessories (scarves, shawls, hats) were constructed, and a photoshoot was arranged to present the objects against a purposely designed backdrop (Fig. 5.12). The accessories were then paired with clothing and displayed for auction in the atrium of Central. The exhibition (featuring objects and photographs from *Strung*) was accessible to students and staff throughout Central's 'Green Week' (enabling people to view the pieces and bid on an item). The project also sparked another project at Central, which was to use discarded fabric offcuts from the costume department to create brightly coloured brooches that we could sell during the school's Green Week. This initiative engaged

Fig. 5.12 Photoshoot of wearable artefacts from *This Is Not Rubbish* Craft Circle. (Photo: Alex Blake)

multiple departments with many of the staff and students wearing the brooches as a symbol of their commitment to environmental issues. From an AUD 30 bag of salami netting material and donated costume offcuts, we raised £338 (roughly AUD 600) that was donated to a community garden for the local housing estate (Dorney Residents Association), giving them a modest start towards funding their project as well as facilitating a connection between the University and the local community.

The reality is that a wide abundance of discarded materials lie waiting for the unique chance of being acknowledged and reinvigorated.

Discovering these remnants, and bringing out their co-creative potential, is often one of the first steps that a scenographer can take towards pursuing sustainable practice. Crucially, this includes considering its post-performance life. A theatre object's further repurposing and value-making adds an additional opportunity for the designer—to reimagine closing night not as an endpoint, but as an activation point for further making, recirculation and re-appreciation.

Ingold (2010, 92) describes makers as 'wanderers, wayfarers, whose skill lies in their ability to find the grain of the world's becoming and to follow its course while bending it to their evolving purpose'. *This Is Not Rubbish* sought out opportunities to activate ecological potential by instigating, sharing, collaborating and engaging with others in the pursuit of a more integrated and ecologically rewarding creative practice. Co-creation and celebration were enacted through engaging local scenographers and audiences in performance-making processes that highlighted the value of reclaimed materials. Circulation led back to co-creation through the repurposing of the material into new artworks that were then further 'celebrated' through an exhibition and auction, allowing the wearable pieces (and subsequent funds) to be 'circulated' into the wider community. *This Is Not Rubbish* highlighted Ecoscenography's circular framework as one that gives value to the entirety of the designer's process.

The Living Stage

Bringing theatre and ecology together to forefront sustainability issues was a major inspiration for *The Living Stage*—an ongoing global initiative that combines stage design, horticulture and community engagement to create recyclable, biodegradable, edible and biodiverse performance spaces. Part theatre, part garden and part food growing demonstration, *The Living Stage* uses the cycles of Ecoscenography to engage people in developing a greater understanding and appreciation of the living world. The co-created community-grown spaces become the setting for performing and celebrating ecological stories, before being circulated back into the communities that helped grow them: 'physical structures become garden beds and community spaces; plants become food; and waste

Fig 5.13 CreateAbility and Born in a Taxi in *Produce* (*The Living Stage*, Castlemaine State Festival, 2013). (Photo: Gisela Beer)

becomes compost' (Beer et al. 2018c, 343). A central focus of *The Living Stage* is to bring a regenerative focus to Ecoscenography that creates opportunities for thrive-ability across more-than-human systems.

The first living stage was created for the 2013 Castlemaine State Festival in Australia and grew out of imagining a new kind of theatrical space—one that was *literally* aligned with ecological systems. Created by the rural community of Castlemaine under the guidance of local permaculturalists (Hamish MacCallum and Sas Allardice), the original project featured an amphitheatre of climbable apple-crate garden walls and portable garden beds, each culturing edible plants (Fig. 5.13). It acted as both a venue and source of inspiration 'for a number of local performance groups whose brief was to create experimental works that drew on the concept of regeneration and interacted with the unique design that surrounded them' (Beer 2015, 2). After the festival, the stage of apple crates and plants was donated to several community gardens for educational projects.

Since making its debut in Castlemaine, the concept has travelled to Cardiff (Wales); Glasgow (Scotland); New York (USA); as well as Armidale, Lorne and Melbourne[2] in Australia. Each Living Stage evolves out of a direct response to the localities of site, ecology and community. No project is ever the same, yet they share clear commonalities: the celebration of multisensory elements, effective and multi-level engagement with audiences, and a legacy that stretches on long after the final performance. At the crux of the project is the notion of community-engaged and 'place-based design processes to foster equity and togetherness on global-to-local issues' (Beer et al. 2018c, 343). With each iteration, *The Living Stage* concept has progressively become more engaged in the desire to enhance the connectivity and integration of more-than-human places in response to climate change, social inequity, food scarcity and biodiversity loss. It is a direct response to 'what can theatre and performance design do?' in the face of increasing environmental concerns.

Integrating cyclic thinking into the design process has been essential to the success of *The Living Stage*. Ecoscenography's cycles of co-creation-celebration-circulation mimic permaculture[3] principles of working with or replicating patterns of the natural world to build ecologically sensitive practices. As well as aligning with ecological systems, *The Living Stage* creates spaces for more-than-human collaboration and communion, where 'artists, ordinary members of communities, and nonhuman beings become enfolded in each other's projects, in each other's lives; they come to need each other in diverse, passionate, corporeal, meaningful ways' (Haraway 2016, 71–72).

Co-creating with Living Systems

In *Through Vegetal Being* (2016), philosophers Luce Irigaray and Michael Marder argue that it is through 'vegetal' (or plant relating) activities that our relations with the more-than-human world can be reignited. Interacting with living plants can provide a meaningful engagement with the living world that respects its presence and agentic capacities—the effects of which can have positive and reciprocal benefits for both human and environmental systems. In Ecoscenography, gardening is seen as co-creative act, which provides an opportunity for designers to work with ecosystems as well as bring people together to strengthen community bonds. This form of collaboration is achieved by enlivening people's connection with nature through embodied experience and provides opportunities for participants to engage in food growing or biodiversity enhancing plantings that are specific to the site. While the nuances of this co-creation experience differ across contexts and communities, each living stage project starts in response to the place and local people of which it is part. The projects use local knowledge as a way of maximising resources (such as the apple crates in Castlemaine or found objects on the streets of Cardiff and New York), as well as demonstrating the benefits of incorporating community engagement into the growing and construction of the stage.

Essential to each project is the collaboration with local plant specialists, horticulturists and permaculturists who provide place-specific expertise that is vital for choosing plants and growing techniques for the specific region. Their detailed knowledge of the area, culture and climate (especially in Australia, where high temperatures and drought often prevail) is crucial to the success of the work. For example, collaborating with permaculturists in Castlemaine, Victoria—a region of dry forests on the cusp of Australia's semi-arid zone—allowed the team to engage with nature's systems to maximise plant-life and food growing productivity without toxic pesticides or chemicals. The permaculture specialists also helped guide and implement sustainable design processes more generally through their focus on waste-free and non-toxic practices.

Even with the guidance of experienced local gardeners, every Living Stage project has been a steep learning curve in co-creating in synchronicity with nature's processes. Integral to this understanding has been the (sometimes harsh) realisation that nature cannot be hurried or slowed down. Instead, I have had to find ways to adapt my design to local conditions and unreliable sources, such as local climatic conditions (e.g. too little or too much sun) or nutrient-poor soil. Ecoscenography embraces serendipitous opportunities that are not beholden to preconceived ideas. Working in tune with ecological processes entails adopting an open, relational and emergent process, which requires an ability 'to respond, adapt to and evolve with change and surprise' (du Plessis 2012, 17). This entails working in synchronicity with nature's processes and allowing unplanned and unexpected elements to extend the atmosphere, beauty, spontaneity and celebration of a work.

Community engagement is central to *The Living Stage*—from planting workshops to harvesting, to celebration and the distribution of the outcomes. Local residents and community groups are invited to plant seedlings and learn about permaculture principles, as well as partake in other components of set and costume construction (Fig. 5.14). In Glasgow (*Uprooted*, Eco Drama, Glasgow, 2015), *The Living Stage* project was used to empower young people to directly experience what it means to care for their environment by engaging four primary schools in the co-creation of an edible set which toured Scotland (Beer 2017) . In Cardiff

Fig. 5.14 Community workshops (*The Living Stage*, Castlemaine State Festival, 2013). (Photo: Gisela Beer)

(*The Trans-Plantable Living Room,* World Stage Design, 2013) the stage was grown by 12 volunteers from a local community garden, with many of them taking a lead on the design, building and hosting of the space. In New York (*The Living Stage NYC,* Superhero Clubhouse, 2017), community drama workshops brought together linguistically and ethnically diverse seniors, fifth graders and neighbouring residents to participate in movement sequences and drama games that occurred adjacent to the design construction on the site (Beer et al. 2018c) (Fig. 5.15). In Armidale (*The Bower Stage,* EcoArts Australis, 2016), local artists led an open-access co-creation process which invited people from local schools, community gardens, disability arts groups and disadvantaged youth organisations to take part in collecting materials from the surrounding area and contribute to the construction and decoration of the dome-like space (Beer et al. 2018b) (Fig. 5.16). In responding to the possibilities of place, every Living Stage project curates opportunities for more-than-human connection, transformation and evolution.

Fig. 5.15 Drama workshops for *The Living Stage NYC* (Superhero Clubhouse, New York). (Photo: Dylan Lopez)

Fig. 5.16 Community workshops for *The Bower Stage* (Eco Arts Australis, Armidale 2016). (Photo: Laszlo Szabo)

Celebrating Thrive-Ability

Recognising the value of stories in activating ecological connection and permeating psycho-social spheres is central to *The Living Stage*. Regenerative development experts Pamela Mang and Bill Reed (2012, 29–30) highlight how stories can play a valuable role in engaging people in understanding ecological complexities, assisting to collectively deepen connections and reimagine futures. Many of the performances and activities that take place on *The Living Stage* work to accentuate the characteristics and dynamics of living systems. Each project offers a platform for local communities and artists to perform themes of regeneration and abundance, and to celebrate thrive-ability. For example, in Armidale, local high school students performed a promenade eco-theatre piece that interacted with the corridor design of suspended native flora, while in Glasgow, a professional theatre show for children connected audiences to nature through the celebration of a weird and topsy-turvy world of vibrant plant-life characters. In Castlemaine, a mixed-ability performance exploring human relationships with the natural world created scenes that celebrated the possibilities of the living design—from nibbling at herbs, climbing up apple-crate walls, cutting wheatgrass and juggling vegetables from the garden (Figs. 5.17 and Fig. 5.18). In all living stages, programming is used to accentuate the design, such as live 'cook offs' using ingredients from the garden set or the opportunity to have a cup of herbal tea whilst seated in the 'living' room design.

Fig. 5.17 On set with Born in a Taxi's Penny Baron (*The Living Stage*, Castlemaine State Festival, 2013). (Photo: Gisela Beer)

Fig. 5.18 Gardening scene with CreateAbility and Born in a Taxi in *Produce* (*The Living Stage*, Castlemaine State Festival, 2013). (Photo: Gisela Beer)

In Armidale, local teachers and theatre practitioners used *The Bower Stage* as their inspiration and performance space to create an ecological-themed theatre production which engaged students from three nearby high schools to devise a theatre piece that explored environmental issues. Together the teachers and students created a vaudeville-style performance in which audiences were invited to journey up a corridor circled by pine trees and native plant sculptures to witness an array of peculiar characters and tableaux. Along the way, spectators were greeted by scenes that included:

> A bearded moss lady who described the health and beauty benefits of moss, two characters in picture frames debating climate change, a scene where characters were fighting over buckets of clean water from the nearby Black Gully creek…an intimate tableaux of violinists playing for a tea party of plants, before inviting audiences to 'nibble' at home-grown lettuce and strawberries. As they left the performance space, spectators visited the Fortune Teller, who gave them a 'personal reading' on a tarot card with an

> environmental message (provided to each person as they entered the space), which focused on the individual responsibility to take environmental action. (Beer et al. 2018b, 31)

A key part of the work was to raise ecological consciousness with young people of the local area—those who will be most impacted by humanity's inaction on environmental issues. The audience included a broad range of participants from many cultural backgrounds, including adults and families with young children in attendance during the day for storytelling in *The Bower Stage*, and a preponderance of teenagers later in the evening that used the space as a 'chill-out room' (Fig. 5.19).

As well as being a space for performance, *The Living Stage* is always designed as a standalone installation—a showcase for collaboratively created design and local sustainable food growing—where people are encouraged to walk amongst the structure and engage with the greenery. Designs are diverse, ranging from an amphitheatre of stacked apple crates

Fig. 5.19 Storytelling for young children and their families in *The Bower Stage* (Eco Arts Australis, Armidale 2016). (Photo: Laszlo Szabo)

Fig. 5.20 Outdoor 'living room' with plants bursting from household objects (*The Trans-Plantable Living Room*, World Stage Design, Cardiff, 2013). (Photo: Mike Medaglia)

filled with edible plants (Castlemaine, 2013), a reclaimed timber platform surrounded by native and edible plants (Lorne, 2018), to outdoor 'living rooms' with plants bursting from household objects (Cardiff, 2013; Glasgow, 2015; New York, 2017) (Figs. 5.20 and 5.21). As Lisa Woynarski (my collaborator on *Trans-Plantable Living Room*) describes:

> Under a large black walnut tree in Bute Park an extensive set of vintage and recycled living room furniture has been installed. The set has been planted with colourful, edible plants, flowers and herbs, all poking up unexpectedly from ordinary household items. Picture frames hang from one of the branches, framing the surrounding park as artworks in the space. A yellow wall clock hangs on the trunk of the tree. A boxy, grey hollowed-out television grows sprouts. Tables made of large wood spools have been planted with edible flowers and woven with willow. Wooden chairs have been fixed to a base where orange and red nasturtiums are blossoming, while the seats of other chairs have been planted up with lush greens. A 'chandelier' of

Fig. 5.21 Outdoor 'living' cabinet displaying images from the community workshops (*The Living Stage NYC*, Superhero Clubhouse, New York, 2017). (Photo: Dylan Lopez)

> yellow and red chilli peppers hangs over the set. Herbs and flowers are growing from dresser drawers, the corners of armchairs and sofas, door frames, coat racks, tea cups, handbags, books and rain boots. Passers-by, dog walkers, childrenc, and people of all ages walking through the park to get from one side of the city to another, stop at the novelty of the installation. They take photos and are invited to sit down and have a cup of tea made with some of herbs grown in the living room. (2020, 143)

Just as *The Trans-Plantable Living Room* offered audiences an immersive multisensory spatial experience, spectators of all Living Stages are invited to touch, smell and 'have a nibble at the stage', as well as pick produce to take home with them at the end of the day (Fig. 5.22). Children, in particular, revel in tasting unusual herbs and flowers, and embrace experiencing new textures, flavours and sensations. This tangible and multisensory engagement with the performance space is a signature

Fig. 5.22 Fresh produce, tea and dried apple rings harvested from the garden (*The Trans-Plantable Living Room*, World Stage Design, Cardiff, 2013). (Photo: Mike Medaglia)

aesthetic, just as every effort is made to ensure that the spaces and performances are free and accessible to people from all walks of life.

In New York, *The Living Stage* was a critical part of advocating for the importance of urban green spaces in large cities. In six weeks the team transformed a large grey asphalt-lined space on the Lower East Side into a lush, playful and green event space (Figs. 5.23 and 5.24) in the middle of a bustling city:

> The home-inspired design included several 'living' rooms, including a perimeter of flowers bursting out of old food drums and a corridor of plant lampshade-sculptures leading to a children's play-space, complete with a blackboard to draw upon, plant-propagated school desks and chairs, old sneakers sprouting greenery and a turf lined school bench. Close to the fence line, a collection of suspended frames bordered the tea room, which included plants growing out of kitchen cabinets, teapots, canisters, baskets and chairs. From here, audiences were led into a living room area of arm-

Fig. 5.23 Melzer Park, view facing the senior centre, with painted lines marking the start of the design intervention (*The Living Stage NYC*, Superhero Clubhouse, New York 2017). (Photo: Dylan Lopez)

chairs and a turf-lined couches, complete with a sewing table displaying knitted grasses and seed buttons. At the centre of the space was a large dining room table, positioned on a raised platform surrounded by apple boxes of flowers and larger shrubs. The park became so alive with greenery and colour, it was almost unrecognisable to some residents. As one audience member remarked, 'it was astonishing to see what was basically a patch of concrete in the middle of the city transformed into a lush and playful garden'; another remarked, 'it was a home, in the way that a home is shared, free and full of conversation'. On the day of the opening, a diversity of seniors and neighbours of all ages filled the park, enthusiastically engaging with the installation and the performances that inhabited them. The program included acts by local musicians, comedians and resident Chinese folk dancers, followed by a salsa music 'dance party' requested by the seniors. The highlight of the day was the presentation of the community play created by Superhero Clubhouse in collaboration with the seniors and school students, which was enacted by a troupe of professional actors for a rapt audience. (Beer et al. 2018c, 354–359)

Fig. 5.24 Overview of the final design as seen from the senior centre (*The Living Stage NYC*, Superhero Clubhouse, New York 2017). (Photo: Dylan Lopez)

The Living Stage NYC demonstrated opportunities for performing intergenerational and cross-community connections that advocated for traditionally underrepresented and diverse BIPOC (Black, Indigenous, and People of Colour) voices on environmental issues. Through fostering collective processes of co-creation and celebration, the project forged relationships across generational, economic, linguistic and ethnic divides. As one senior exclaimed, 'it was like we were brothers and sisters sharing the same space, despite different languages, cultures and creeds' (Ibid.). The very act of making and performing places together in real time and space provides opportunities for collective memories (Coghlan et al. 2017, 79). These collective memories can act as powerful agents of change (Mang and Reed 2012, 29–30), permeating psycho-social spheres and influencing the lives of those connected to them (Bal and van Boheemen 2009). By using Ecoscenography as a tool to envision alternative futures, *The Living Stage NYC* offered a glimpse into how equitable and thriving environmental, social and cultural gathering places might be made possible (Fig. 5.25).

Fig. 5.25 Community atmosphere during performance (*The Living Stage NYC*, Superhero Clubhouse, New York 2017). (Photo: Milton Perks)

At the heart of *The Living Stage* is the investigation of alternative narratives for (re)connecting urban communities to ecological themes and practices. Sustainability issues are intrinsically social, and any hope of changing people's hearts and minds requires breaking down cultural and social boundaries that reinforce nature-human dichotomies. The coming together of local and diverse communities to 'celebrate place' is an important component of Ecoscenography. Living Stage projects provide an opportunity to think creatively about environmental issues, outside the box of political rhetoric and partisan ideals—to celebrate the intrinsic value of our socio-ecological togetherness. *The Living Stage* has had the capacity to engage people of all walks of life in environmental projects—people who might not immediately identify themselves as being devoted to sustainability, but through the act of being involved in the project may find themselves suddenly entwined in the process. Here, the notion of 'celebration' is seen as one that actively fosters reciprocal connection of more-than-human relationships through the hands-on co-creation of a thriving future. These components play an active role in inspiring hearts and minds, to empower communities to take action in building resilient communities.

Circulating Positive Legacies

Co-designing places, celebrating together and circulating rewards are essential components of *The Living Stage* legacy. Creating a positive footprint that contributes back to the local environment and community is an ambitious target that is achieved by enlivening people's connection with nature, fostering gardening and communication skills, as well as producing physical artefacts (e.g. portable gardens or furniture items) that can be utilised by the communities for many years after the show. In Castlemaine, the 14 apple crates (filled with edible plants) that made up the stage design were donated to a local TAFE centre to facilitate food growing projects and educational initiatives in otherwise underutilised spaces (Beer 2015) (Fig. 5.26). After our final showing in Glasgow, we turned to the task of transplanting and installing the stage as a permanent feature in our partner school's playground, turning an ugly metal fence into a beautiful space for future gardening and storytelling (Beer 2017). In New York, we opened up an informal stall in Melzer Park to distribute furniture pieces and plants to senior residents and neighbours (Beer et al. 2018c, 354–359). In Lorne, our timber stage platform was donated to a local high school for student performances, while the plants were distributed out to the community garden, school and land management authority for further planting.

In Cardiff, furniture and plants were integrated back into the community garden: the door became part of the fencing, the teapots and cups supplying colourful and much-needed crockery for the volunteers' tea drinking ritual, the lounge chairs were installed in the existing shelter where they provided a comfortable social space for ongoing communal gatherings and newly initiated 'garden social nights'. The Cardiff project had a clear social impact on the lives of those involved, boosting the

Fig. 5.26 Apple crates (filled with edible plants) from the set design were donated to a local TAFE centre to facilitate food growing projects and educational initiatives (*The Living Stage*, Castlemaine State Festival, 2013). (Photo: Tanja Beer)

confidence and sociability of a number of the community garden volunteers. The project demonstrated how engaging communities in the creative possibilities of growing food could also have benefits beyond the growth of the garden. Participants were able to engage in ways that supported physical, psychological and social health as well as exposing creative avenues that may have otherwise had little chance of being explored.

The Living Stage aims to demonstrate the potential of Ecoscenography's circular framework through developing creative and meaningful connections, partnerships, and ecological and cultural networks. Co-creation, celebration and circulation are enacted by taking a uniquely local approach to designing with and for place, where community members and audiences become a crucial part of the aesthetic experience. Circulation is one of giving back to the community, where scenographic elements are designed with contribution in mind. Fundamental to the development of Ecoscenography is the testing of design practices that are capable of contributing 'positive, mutually reinforcing, enduring benefits to human and ecological systems' (Robinson and Cole 2015, 4). For Kathleen Irwin (2007, 54), scenography can be 'a way of giving back places to people and restoring, however temporarily, a sense of the reciprocal relationship between who we are and where we are'. The work in this chapter demonstrates a similar focus. One in which the ultimate goal of Ecoscenographic practice is engaging with regenerative practice and seeking out projects that have the capacity to contribute to the socio-ecological health of place. While it is difficult to quantify these effects, projects like *The Living Stage* demonstrate how designers and theatre makers have the potential to shift narratives and perceptions of place—a process that can rapidly catalyse engagement, cultivate empathy, precipitate action and generate hope.

Beyond *The Living Stage*

A key consideration of Ecoscenography is what it offers within the hybrid terrain of transdisciplinary, spatial, social and ecological discourse. How might projects like *The Living Stage* be expanded further to engage with a diversity of fields, such as climate science, biosciences, built environments or urban planning? Such engagements point to a spectrum of opportunities for the field of Ecoscenography, including a consideration of how performances might be integrated with the natural sciences and natural systems, to encourage empathy for, and an understanding of, ecosystem processes. Indeed, my latest projects have increasingly taken me outside of the theatre, into the expanded realm where Ecoscenography's more-than-human perspectives, performative and narrative-based strategies intersect with environmental psychology, placemaking and science communication.

Running Wild (Polyglot Theatre, Melbourne, 2016) was one such project. *Running Wild* was a practice-led research Initiative[4] that aimed to explore new approaches to environmental learning, approaches that celebrate children's agency and support them to co-create their own futures.

Fig 5.27 Spirit animal costumes made by the children using reclaimed materials (*Running Wild*, Polyglot Theatre, 2016). (Photo: Sarah Walker)

The project introduced nature-deprived primary school students[5] to their local nature reserve through an amalgamation of Ecoscenographic activities, including the making of spirit animal costumes from reclaimed materials (Fig. 5.27), and creation of an outdoor promenade of 'bush cubbies' and performances for the children's families. A qualitative study run in parallel with *Running Wild* demonstrated increased nature connection amongst participants. Importantly, however, the study also confirmed the importance of nature-based costume-making, set-building and performing to improving learning capacities and wellbeing (Beer et al. 2018a). A key finding was the discovery of costume and cubby-making as a noteworthy device for 'becoming animal' which allowed children to enact exercises in freedom, to seek out 'wildness' and confront the unknown (Cooper Marcus 1978; Derr 2006) as well as increase empathy for more-than-human:

> A significant component of *Running Wild* was engaging the children in the experience of making cubbies ('artistic shelters' or 'habitats') for themselves. While we experienced some initial challenges in the children's lack of fine motor skills, designing cubbies was a high point for many of the group. As the children spent time exploring and fossicking for branches and leaves for the shelters, we witnessed how the act of 'making' became an important part of finding their own place in nature. As one child remarked, "I remember how we got to make cubbies and have lots of fun and be able to just do things freely and be able to like, imagine things and trying to put it into reality". (Beer et al. 2018a, 13–14)

Running Wild revealed the value of Ecoscenographic tools in opening up new opportunities for young people to experience their own agency and capacity for contribution by engaging in their local environments.

Projects like *The Living Stage* and *Running Wild* demonstrate ways through which Ecoscenography can engage urban communities in highly visual, sensorial, interactive and participatory events that can turn complex ecological ideas into tangible concepts and active learning processes. I contend that Ecoscenography has the potential for creating new public engagement tools and strategies that can examine ways in which creative practice can generate ecosystem restoration and community vitalisation. In collaboration with community partners, an Ecoscenographic project might engage with 'stories of place' through performative, dramaturgical, narrative and experiential site-based approaches. Considerations might include how urban spaces have transformed past ecosystems, how historic and contemporary landscapes intersect with human trajectories and spatial hierarchies, and how these stories might be revealed to audiences through new forms of communication.

Ecoscenographic Futures

This Is Not Rubbish and *The Living Stage* were test cases in Ecoscenography: they sought to explore, and demonstrate, how performance design can implement a regenerative framework to activate socio-ecological potential. In both cases, I stepped outside traditional theatre spaces to pursue these ideas, but I am acutely aware that such projects could alienate some theatre makers. The concern becomes one of exclusivity, which only reaffirms the binaries this book is trying to disrupt. Indeed, the idea that Ecoscenography can only exist in expanded, experimental and/or community-engaged practices undercuts its potential. Thus, it is imperative that Ecoscenography is tested and explored across an array of platforms and possibilities. For the paradigm to cut through, theatre makers (including myself) must gain an understanding of the challenges and opportunities for implementing socio-ecological strategies within a range of performance spaces and cultural venues, and refine theories and practices accordingly. A guiding principle, highlighted in Chap. 4, must be 'radical ideas that are not rushed'.

This sentiment is echoed in Rosie Elnile's *Prayer* (The Gate Theatre, London, 2020), a ground-breaking online scenographic work that explored what it might mean to grow a garden in London's Gate Theatre. Staged during Britain's first COVID-19 lockdown, Elnile invited audiences across the globe to tune into a digital-visual-aural exploration of her design thinking, which interrogated white, colonial assumptions of theatre, space, politics and power. In viewing the digital provocation, audiences could take their time to absorb a series of precedents and readings along with beautifully designed graphics, GIFs and audio files that explored ideas of nurturing a garden on The Gate stage, while also questioning spatial politics. The work executed a novel approach in the use of asynchronous online and digital platforms to explore scenographer-led processes and aesthetics that re-envision what theatre can be in an age of social and environmental advocacy.

In an accompanying essay, Elnile (2020) highlights how designing for theatre (no matter the context) is an inherently political act, one where scenographers can 'test out, in small ways, the most radical or hopeful version of how a space might operate…to try to imagine politics that feel impossible'. To do this, Elnile argues that designers should see theatre spaces not as inert, but as 'active site[s] for change' that have their roots in the socio-ecological-political facets of place (Ibid.). This also means interrogating traditional ways in which scenography and theatre making are taught:

> As a designer, it is easy to see these spaces not in relation to geography or history but as a blank canvas: where the sets we design exist only in relation to the text, actors, sound, lighting elements and the architecture of the building. When I studied set design, the majority of my syllabus was focused on designing speculative projects for theatres such as the Almeida and the Hackney Empire…It was required that we do weeks of careful and detailed research into the time period and geography of, say, Chekhov—but no time or attention was given to the real geography of the space or the real cost of the design. Who are the community that live near the theatre? What is the history of that community? What is the colonial or neo-colonial history of that building? What impact is that history having in this moment? What is the ecological cost of our proposed design, where might those materials come from and how are they manufactured? Why is Chekhov's life and community and aesthetic given more attention than the actual spaces that we are making in and the people we are making for? (Ibid.)

Elnile's essay speaks to the notion that spaces *are* places, places filled with histories, hierarchies and subjectivities. There is no such thing as an 'empty stage', and part of our job as designers is to acknowledge and speak to these tensions and truths, to look beneath the layers. As part of her radical reimagining of performance design, Elnile calls for slower, more respectful and 'horizontal methods of making' that allow designers and communities to be at the centre of the process—one that creates opportunities for new voices and perspectives. As she contends:

> It's impossible to imagine that we can reduce the ecological cost of theatre making without designers and visual artists being in the core creative team of a theatre building. We need space, time and power to imagine new aesthetics and processes. We need to be involved in programming. With funding and time, it is possible to reimagine design as a practice that is less extractive and focused on specificity and care. Set design could be a laboratory for new ways of thinking about the politics of space and a tool to imagine new futures. (Ibid.)

Elnile's provocation represents a timely voice in rethinking how we make theatre that goes far beyond superficial engagement with socio-ecological issues. Instead, her vision is one that tests new ways of doing that embrace wider perspectives and post-colonial worldviews for the future of our field. My hope is that Ecoscenography represents a pathway for such a vision. As demonstrated by the above projects, and many other examples in this book, a new wave of theatre making is emerging that is more ecologically, socially and politically conscious, inclusive, collaborative and transdisciplinary. Theatre artists like Rosie Elnile are recognising that they can utilise the highly visible position of the performing arts as a platform to confront and reimagine societal values, ethics and biases.

This reimagining is further discussed in Rachel Hann's Afterword and provides a launchpad for much-needed further thinking beyond this book. As Elnile and Hann's words reveal, there are a great many avenues for further practice and research that speak to the provocation of 'what theatre can do' in an age of increasing ecological crisis.

Notes

1. The original intention of the salami netting material was unknown to me, and it was only discovered shortly prior to the first performance. The fact that the netting had been initially intended for the production of making sausages did not play a role in my creative investigation of the material. For me, not knowing the object's original purpose was advantageous as it opened up avenues for exploration—to engage with its agency and vitality: 'to think from the materials, not about them' (Ingold 2012, 437). In this case, the material was not judged by the knowledge of what it was or what it should do but by the ability to engage with what it could do.
2. The Melbourne project, entitled *The Living Pavilion*, was a First Nations-led, transdisciplinary project connecting Indigenous knowledge, ecological science and sustainable design through participatory arts (Beer et al. 2019). The event took place between 1–17 May 2019 and was held at the University of Melbourne's Parkville Campus (Victoria, Australia) on unceded lands of the Wurundjeri peoples of the Woi Wurrung language group, who have belonged to and been custodians of those lands for more than 65,000 years. *The Living Pavilion* was co-produced by Cathy Oke, Tanja Beer and Barkandji woman Zena Cumpston as part of the ART+CLIMATE=CHANGE festival (23 April–19 May 2019). More

information about the project can be found via: https://nespurban.edu.au/wp-content/uploads/2020/02/The-Living-Pavilion-Report.pdf.
3. Permaculture is a waste- and chemical-free discipline that employs local composting, recycling, seasonal and companion planting, natural sunlight and rain or tank water to grow food (see Mollison 1991; Holmgren 2003).
4. *Running Wild* was a collaboration between the University of Melbourne, Polyglot Theatre, Royal Botanic Gardens Cranbourne, Parks Victoria, Linda Mullett (a Bunurong Elder from the Mornington Peninsula area) and Mahogany Rise Primary School (North Frankston). The project engaged 36 Grade six primary school students in their local ecology—the Pines Flora and Fauna Reserve (the children's local 108 hectare bushland) and the Royal Botanic Gardens in nearby Cranbourne—through participatory arts, science and Indigenous culture over a period of 12 days across five weeks.
5. The local school involved in the research (Mahogany Rise Primary School) struggles with associated issues of school absenteeism, poor school achievement and other signs of disengagement with mainstream, traditional modes of learning. Children's time outside of school is largely unsupervised and is spent playing video games or watching television rather than playing outside. This context reflects trends in declining independent play, increasing sedentary and indoor play experience and growing ecophobia amongst young people.

References

Bal, Mieke, and Christine van Boheemen. 2009. *Narratology: Introduction to the Theory of Narrative*. Toronto, Canada: University of Toronto Press.

Barad, Karen. 2003. Posthumanist Performativity: Toward an Understanding of How Matter Comes to Matter. *Gender and Science: New Issues* 28 (3): 801–831.

Beer, Tanja. 2015. The Living Stage: A Case Study in Ecoscenography. *Etudes* 1 (1): 1–16.

———. 2016a. Ecomaterialism in Scenography. *Theatre and Performance Design* 2 (1–2): 161–172.

———. 2016b. Reimagining the Ruins of Scenography. *ASAP Journal (Association for the Study of the Arts of the Present)* 1 (3): 487–511.

———. 2017. The Boy and the Sunflower: The Role of Theatre in Communicating Ecological Processes Through the Creation of Living Stages.

In *Building Sustainability with the Arts: Proceedings of the 2nd National EcoArts Australis Conference*, ed. David Curtis. Newcastle upon Tyne, UK: Cambridge Scholars Publishing.

———. 2018. Saved from the Scrapheap: Revealing the Creative and Ecological Potential of Societal Leftovers in Scenography. *Performance Research* 22 (8): 107–114.

Beer, Tanja, Andrea Cook, and Kate Kantor. 2018a. Running Wild: Engaging and Empowering Future Custodians of Place Through Creative Nature-Based Play. *Journal of Public Pedagogies* 3: 5–19.

Beer, Tanja, David Curtis, and Julie Collins. 2018b. Chapter 3 Innovation: Creativity as a Renewable Resource for the Eco-City. In *Enabling Eco-Cities: Defining, Planning, and Creating a More Thriving Future*, ed. Dominique Hes and Josh Bush. London, UK: Palgrave Pivot.

Beer, Tanja, Lanxing Fu and Cristina Hernández-Santín. 2018c. Scenographer as Placemaker: Co-creating Communities Through The Living Stage NYC. *Theatre and Performance Design* 4 (4): 342–363.

Beer, Tanja, Cristina Hernandez-Santin, Zena Cumpston, Luis Mata, Rimi Khan, Kirsten Parris, Christina Renowden, Rachel Lampolski, Dominique Hes, and Blythe Vogel. 2019. *In The Living Pavilion Research Report*. Victoria, Australia: The University of Melbourne.

Bennett, Jane. 2010. *Vibrant Matter: The Political Ecology of Things*. London, UK: Duke University Press.

Coghlan, Alexandra, Bev Sparks, Wei Liu, and Mike Winlaw. 2017. Reconnecting with Place Through Events: Collaborating with Precinct Managers in the Placemaking Agenda. *International Journal of Event and Festival Management* 8 (1): 66–83.

Cooper Marcus, Clare. 1978. Remembrance of Landscape Past. *Landscape* 22 (3): 35–43.

Deleuze, Gilles, and Felix Guattari. 2004. *A Thousand Plateaus: Capitalism and Schizophrenia*. Minneapolis, USA: University of Minnesota Press.

Derr, Tori. 2006. Sometimes a Bird Sounds Like a Fish: Perspectives on Children's Place Experience. In *Children and Their Environments: Learning, Using and Designing Spaces*, ed. Christopher Spencer and Mark Blades, 108–123. Cambridge, UK: Cambridge University Press.

Elnile, Rosie. 2020. Design is a Political Act: Let's Use it to Reshape the Future. In *Gate Theatre*. https://www.gatetheatre.co.uk/blog/design-is-a-political-act-lets-use-it-to-reshape-the-future-by-rosie-elnile/.

Haggard, Ben. 2002. Green to the Power of Three. *Environmental Design and Construction* 5 (2): 24–31.
Haraway, Donna. 2016. *Staying with the Trouble*. Durham, NC: Duke University Press.
Holmgren, David. 2003. *Permaculture: Principles & Pathways Beyond Sustainability*. Castlemaine, Australia: Holmgren Design Services.
Ingold, Tim. 2010. The Textility of Making. *Cambridge Journal of Economics* 34 (1): 91–102.
———. 2011. Introduction. In *Redrawing Anthropology: Materials, Movements, Lines*, ed. Tim Ingold, 1–21. Farnham, UK: Ashgate Publishing.
———. 2012. Toward an Ecology of Materials. *Annual Review of Anthropology* 41 (1): 427–442.
Irigaray, Luce, and Michael Marder. 2016. *Through Vegetal Being: Two Philosophical Perspectives*. New York, USA: Columbia University Press.
Irwin, Kathleen. 2007. *The Ambit of Performativity. How Site Makes Meaning in Site-Specific Performance*. Helsinki, Finland: University of Arts and Design Helsinki.
Kompanek, Christopher. 2012. For Donyale Werle, It's Easy Being Green. In *American Theatre: A Publication of Theatre Communications Group*. https://www.americantheatre.org/2012/09/01/for-donyale-werle-its-easy-being-green/.
Mang, Pamela, and Bill Reed. 2012. Designing from Place: A Regenerative Framework and Methodology. *Building Research & Information* 40 (1): 23–38.
Miller, Daniel. 1995. *Acknowledging Consumption: A Review of New Studies*. London, UK: Routledge.
Mollison, Bill. 1991. *Introduction to Permaculture*. Tasmania, Australia: Tagari Publications.
Parsons, Liz. 2008. Thompsons' Rubbish Theory: Exploring the Practices of Value Creation. *Advances in Consumer Research – European Conference Proceedings* 8: 390–393.
du Plessis, Chrisna. 2012. Towards a Regenerative Paradigm for the Built Environment. *Building Research & Information* 40 (1): 7–22.
Robinson, John, and Raymond Cole. 2015. Theoretical Underpinnings of Regenerative Sustainability. *Building Research & Information* 43 (2): 133–143.
Woynarski, Lisa. 2020. *Ecodramaturgies: Theatre, Performance and Climate Change*. Cham, Switzerland: Palgrave Macmillan.

6

Conclusion

T. Beer, *Ecoscenography*, https://doi.org/10.1007/978-981-16-7178-4_6

Ecoscenography builds on contemporary notions of theatre and performance design to consider what an ecological approach to scenography *does*—how it affects our ways of thinking and doing within and beyond the performing arts. A central aim of this book has been to demonstrate that embracing sustainability in performance design opens up new modes of relating and collaborating: to work co-creatively with communities, environments, materials and places; to appreciate and advocate for 'more-than-human doing'; to engage in 'acts of care' that foster a respectful and reciprocal communion with the world. This 'permeability of relating' in Ecoscenography seeks to dissolve perceived boundaries between artists, materials, audiences and the broader ecosystem, opening up greater eco-creative responsibilities and considerations for how we conduct our practice and communicate through it.

Ecoscenography's multi-scalar and relational approach calls into question existing hierarchies and power structures that separate and distance the individual from the more-than-human world. Ecoscenography is wary of exclusive approaches to theatre making that are at odds with 'intersectional ecological thinking' (Woynarski 2020, 215). Instead, an ecological approach to scenography campaigns for a shared approach to the democratisation and activation of spaces and places, one that forefronts intersectional community participation and belonging, with the goal of creating social-ecological sites for celebration. This includes problematising and synthesising how places 'are felt in relation to human-centric power geometries (nations, economics, belonging) and the ecological connectivity of non-human assemblages (human-world binaries, ecological relations, process philosophies)' (Hann 2020, 21).

The act of partnering with place is vital to Ecoscenography. Expanded ideas of material co-entanglement promote new considerations of performance design, where ecological transformation ultimately begins with how we respond to where we are, as a primary approach. Ecoscenography cannot exist without its relationship to, acknowledgement of and respect for place, in all its layers of diversity, complexity and nuance. The three stages of Ecoscenography—co-creation (pre-production), celebration (production) and circulation (post-production)—provide a fundamental framework for how holistic, interconnected and symbiotic processes can be integrated into all areas of theatre production. The challenge and the opportunity for the designer are to seek out these relational possibilities as a central premise of their practice.

This book has also sought to showcase the diversity of theatre artists exploring what it means to bring an ecological ethic into their practices. We are standing at the precipice of a burgeoning movement, one that has the potential to drastically change the way we make theatre in the years to come. As I write this, there are arguably hundreds (possibly thousands) of scenographers, performers, directors, stage managers, set builders, costume makers and producers that are integrating sustainability into their practices. These practitioners are slowly influencing policies, production processes and construction methods around the world. Many of their ecological acts may not be widely publicised, but they are happening, in rehearsal rooms, scenic workshops, artist studios, warehouses, basements, online platforms, in communities and schools. They are happening across multiple linguistic and ethnic realms, sometimes behind closed doors, sometimes in public. They are smaller or larger acts of protest; they are the local and often unsung heroes of this movement. This book has only touched the surface of what is out there.

As a researcher, I am primed to look for commonalities—patterns, threads, themes, trends—that might define this moment of ecological awakening in theatre making. What is the defining characteristic of now? Indeed, many an artist, scholar or critic has grappled with (and continues to grapple with) this fundamental question. Perversely, it is often not until we have gained some future perspective that we are able to fully grasp what is happening today. And this is from where I write, from a place of murky observance and experience, rather than objective clarity, in understanding what the recent swell in environmentally engaged scenography and theatre making means for our field.

Despite the short trajectory, there are patterns emerging, ways of working that can be seen across regions and cultures that are changing the way in which we make theatre. Many of these characteristics are evident in previous chapters—practices that forefront ecological values by supporting circular processes and bringing communities together around environmental issues in unique ways. Theatre makers are engaging in a way that reconsiders our abiotic, biotic and technotic relationships with the world, while also acknowledging their intrinsic porosity and interrelation. This includes starting to adopt an expanded view of aesthetics that acknowledges the 'unseen' effects of making spaces—that which may not be immediately evident in the making of the work but has causational potential to form a trans-corporeal by-product of the 'visible' and 'experienced' (such as adding to toxic waste, environmental pollution and child exploitation). This multifaceted and complex approach to aesthetics is already shifting how many scenographers engage mentally, as well as practically, with the issue.

The artists and organisations highlighted in this book have been chosen for their ingenuity in challenging traditional and unsustainable practices across a diversity of backgrounds, as well as conventional and expanded platforms. Many have been at the forefront of integrating ecological values into their productions, well before sustainability in theatre production was a topic of importance and consideration. They are trailblazers, innovators and pioneers who come from an authentic place of ecological concern, which also fuels their creativity. Several of these artists have taken great artistic and financial risks to make their work count in a brutally competitive industry that frequently favours the 'ego' over the 'eco'. Some are scenographers that have increasingly become artists in their own right, stepping away from the mainstream, to have the freedom to explore an ecological approach to what they do, while others have endured the grit and hard work in changing practices within highly traditional theatre venues. Collectively, their work ranges from sustainable set designs on New York's Broadway to wearable plant-costumes in São Paulo and kinetic energy dance performances in Paris. They include performative feasts in Singapore, agricultural theatre festivals in India and community-weaving-as-scenography in Mexico. Few are recognised for the ground-breaking work that they do. Yet, they have significantly informed and inspired my concept of Ecoscenography over the last decade. I am lucky to have had the pleasure of knowing many of these artists, by either collaborating with them directly, or physically experiencing their work, or admiring their work from afar. For me, they represent the variety of Ecoscenographic projects that are already happening across the world and a hint at the future of the field, of what is yet to come.

Steps Forward

In this book I have argued for the necessity for the performing arts to embrace the creative potential of an ecologically conscious future. An integral part of this will be to foster an educational model where sustainability is integrated into theatre pedagogy. While architecture, interior design, product design and fashion have been including sustainability as part of their course curriculum for decades, performance design is a laggard. This must change, and the next generation of designers are already holding theatre educators accountable for their lack of engagement on topics of social-ecological justice and sustainability.

There is no doubt that regulation is needed to progress the agenda of environmentally responsible practices, but we can start rehearsing now. What matters is taking action, and doing so together, in the belief that public organisations and other individuals will follow. Theatre makers must seize the opportunities currently available in their own environments, to encourage others and inspire communities of action by doing. As Anne Justine D'Zmura writes in *Readings in Performance and Ecology* (2012, 178):

> We do not have to wait for others to take action. Theatre artists and educators not only have the capacity for imagining change in our world, we have the tools, the creativity, and the holistic approach demanded by our art to carve out life-changing and environmentally minded experiences for each other and members of the larger community we serve.

The challenge of the ecologically minded scenographer will be to guide new ways of interacting with the living world through spatial, performative, participatory and celebratory processes. Embracing Ecoscenography for a thriving future will require a strong desire to seek out meaningful relationships, opportunities and abundance with a readiness to respond, adapt and evolve to changing circumstances. Helping to create a space for this to occur will also be a key aspect of moving towards a regenerative future.

Engaging with Ecoscenography is not intended as a single pathway; however, this book has provided a framework for how ecological thinking and practice may be applied to scenography. In seizing the opportunities of ecological design, we must continue to evolve our tools, guidelines and road maps in response to a rapidly changing and uncertain world. It is my hope that the paradigm of Ecoscenography assists designers in this process—to reimagine and cultivate stronger relationships with materials, places, communities and ecosystems, and to invest directly in their future.

Ideas of what constitutes an ecological scenography will change over time. The idea that notions of sustainability are constantly changing and growing is a fundamental. The world is not fixed, 'but an ever-fluctuating process of change and transformation brought about through interactions between entities and their responses to changes in their environment' (Hes and du Plessis 2014, 203). As Summer Banks (2014) writes:

> It's clear that there is no single path to creating an environmentally-friendly theatre culture. At the same time, ecological impact is an issue that the theatre cannot afford to ignore. And, as a live art form, performance is uniquely suited to cut through the apathy caused by checklists, endless sensational television specials, or didactic speeches about the coming environmental apocalypse. And this message-in-the-moment isn't just created by texts that address climate change, or by switching to solar-generated electricity and high-efficiency lights. It's created by incorporating these techniques into the production's aesthetics. It is not enough to just talk about them, or even implement them behind the scenes; these concepts also have to be shown. It's long been said that some limitations promote creativity instead of stifling it. So why not approach the "restrictions" imposed by environmentally-friendly strategies as inspiration for innovation, as the very material for performance? The environmental movement is creating techniques and tools that are not only relevant for their green impact, but also for their theatrical potential.

The sentiment is put poetically and powerfully by Alison Tickell (2012) of Julie's Bicycle, and it is with her words that I will close: 'creativity is the most sustainable and renewable energy source on the planet. Let's use it.'

References

Banks, Summer. 2014. Riding a Bike on Stage: Green Theater and Katie Mitchell's Production of Lungs. *Howlround Theatre Commons*. https://howlround.com/riding-bike-stage.

D'Zmura, Anne Justine. 2012. Devising Green Piece: A Holistic Pedagogy for Artists and Educators. In *Readings in Performance and Ecology*, ed. Wendy Arons and Theresa May, 169–179. New York, USA: Palgrave Macmillan.

Hann, Rachel. 2020. Justice Scenographics: Preparing for civilization change in a time of "Anywheres" and "Somewheres". In *Spatial Justice 2.0: A Frame for Reclaiming Our Rights to Be, Thrive, Express and Connect*, ed. Design Studio for Social Intervention (ds4si), 18–23. https://www.ds4si.org/blog/2012/1/23/spatial-justice-a-frame-for-reclaiming-our-rights-to-be-thri.html#:~:text=ds4si-,Spatial%20Justice%3A%20A%20Frame%20for%20Reclaiming%20our%20Rights,Be%2C%20Thrive%2C%20Connect%20and%20Express&text=Space%20is%20currently%20functioning%20as,can%20use%20it%20even%20better.

Hes, Dominique, and Chrisna du Plessis. 2014. *Designing for Hope: Pathways to Regenerative Sustainability*. Hoboken, USA: Taylor and Francis.

Tickell, Alison. 2012. Sustainability Should be at the Heart of Our Collective Artistic Vision. *The Guardian: Cultural Professionals Network*. http://www.theguardian.com/culture-professionals-network/culture-professionals-blog/2012/oct/25/sustainability-arts-council-julies-bicycle.

Woynarski, Lisa. 2020. *Ecodramaturgies: Theatre, Performance and Climate Change*. Cham, Switzerland: Palgrave Macmillan.

Afterword: How to Celebrate

Rachel Hann

There are many timely and urgent ideas featured in this book. Yet, it is the provocation that production is reimagined as 'celebration' that excites. To practise performance as a celebration of human and nonhuman thriving is game changing. To make climate-changed theatre is to celebrate: to celebrate social interaction; to celebrate the deep pasts and deep futures

T. Beer, *Ecoscenography*, https://doi.org/10.1007/978-981-16-7178-4

of performance things and architectures; to celebrate the unique moment in history at which these people and materials have come together; to celebrate the release of energy, whether light from a lantern or the movements of the land. The cyclical model of Ecoscenography and, particularly production as celebration, offers a radical renewal on what it means to design events.

The question I now ask is not why ecologically grounded scenographic practice should be embraced, but rather how. How can theatre and performance celebrate the interweaving of humans, things, and places? By way of an Afterword for this book and as a provocation to future practitioners, I offer three replies to this question:

1. Empty space is a colonial concept.
2. Health is a criterion for scenographic practice.
3. Design for the feelings that have no name.

Each of these is offered as wayfinders to help negotiate the complexities of living with a climate-changed world. Yet, also to embrace the next fifty years as a period of unpresented creativity, of experiment, and of celebrating human interdependence with worlds that sustain, care, and weather uncertain futures.

Empty space is a colonial concept. First and foremost, ecological revolutions will be bound to decolonial practices and philosophies. Global North philosophies of world are predicated on continual growth from 'nothing'. All of the leading theories of ecological design plotted in this book observe that resources are not unlimited. The belief that continual economic growth and human expansion, even to other planets, puts significant weight on technological saviours that may be unrealistic or undesirable. Yet, expansion has been the *raison d'être* for the civilisations that emerged in Europe and Northern Asia. Histories of colonialism are histories of 'discovering' seemingly empty spaces. If they were not empty, there would be a focused and prolonged effort to empty those spaces of pre-existing peoples, histories, and philosophies. To clean them. To cleanse them. The imagination that spaces can be cleansed gives political permission to acts of colonialism. Places are emptied to make way for new civilizations that mark themselves as superior and progressive. The celebration of people and things must challenge these implied permissions to start afresh without considerations for pre-existing ideas, peoples, and materials.

In relationship to scenography, the idea that designers start with an empty void is a manifestation of Global North imaginaries of empty space. The black box model is the epitome of this approach. Where as with the black box studios synonymous with experimental twentieth-century theatre, the black box recedes away from view to offer a space clean of feeling, politics, and history. To start afresh each time. Yet, this is never the case. Instead, scenographies always emerge *with* the geographies, peoples, and histories that intermingle during the act of performance. A decolonial scenography is inclusive of celebrating the local, whether peoples or materials, and the reuse of materials that audiences have seen before. Take pride in offering creative transformations that allow materials to surprise audiences again and again. While in previous cultures of production this might be dismissed as derivative or 'lazy', the continual reusage of locally sourced materials acts as a reminder that scenographies transform feelings of stuff. Make room to celebrate local materialities that afford recognition (from audiences) and reusage (for carbon footprints).

Health is a criterion for scenographic practice. There is one well-worn question that all theatre makers have asked themselves at some point: 'does this look right?' Vision-centric cultures of performance have produced criteria that privilege certain aesthetics, of newness, clean lines, virgin materials, and so on. Yet provocations such as Ecoscenography ask us to focus on not only what scenography *does* but what it *feels* like. What if rather than asking about the look of a production, the question was 'does this place feel healthy?' Health as a criterion for scenographic practice is multifaceted. As outlined by Beer, biophilic design proposes that criteria concerned with serving the needs of the play are reimagined to the health and wellbeing of all enclosed. This shift in thinking does not preclude traditional methods of text analysis. Rather it elevates the findings of these techniques to a broader ecological concern. To generate healthy performance places is to celebrate how and why plays generate feeling: from the temperature of the auditorium to the tensions between narrative and materiality. Scenographies of feeling of scenography are bound to dramaturgies of feeling. Healthy scenographies are scenographies that conceptually irritate (by rubbing up against normative feelings of place) and yet practically care for the psychological, physical, and social wellbeing of the peoples and things it brings together. Beer's model of Ecoscenography is a model for caring beyond strictly human needs, and yet this ethics of care also seeds the conditions for human thriving in a climate-changed world.

Design for the feelings that have no name. The speculative drag artist Victoria Sin offers a similar provocation in reply to their active resistance to cultures of categorisation, whether of gender, art, or meaning. In a period of climate crisis, the need to practise differently is also to change the frames of reference that sustained production cultures of waste. To design for the feelings that have no name is to design for a period of human civilisation that has no precursor. Renewed feelings of world, nonhuman interdependence, and future require emotional experiences that potentially exceed categorisations such as happy, sad, anxious, or hopeful. The seeding of future emotional orientations is undoubtedly squarely within the remit of theatre and performance. To feel with places, to experience a blurring of bodies and environments, is critical for future emotional states that realise Bill Reed's ultimate goal of Regenesis: 'Be Nature'. Ecoscenography sits at this interface. Create events that invite new feelings, new orientations, and new emotions of place. Do not be frightened to celebrate dwelling, voicing, and practising the hitherto politically impossible act of withdrawing from unhealthy cultures of productions without futures. Design for the productions that are yet to come!

To conclude and energise, I leave with a message to future theatre and performance makers: Do not make theatre like we did. You live in a new ecology. Unabashed freedom of creativity (of resource, material, and travel) is historic. Thrive within an ecological limit. Make theatre like you can.

Project Credits (Chapter 5)

First and foremost, I would like to express my deep gratitude to the Traditional Owners of the many Countries that make up Australia, and the various Nations and First Peoples across the world upon whose lands I have had the benefit of conducting my creative projects. I am grateful for the vast array of collaborators and volunteers who took part on *This Is*

T. Beer, *Ecoscenography*, https://doi.org/10.1007/978-981-16-7178-4

Not Rubbish (Strung), The Living Stage and *Running Wild*. Thank you for all your incredible commitment and enthusiasm for making these projects come to life. While the following outlines the numerous people that contributed to my work in Chapter 5, I am also grateful for the many audiences and participants who may not be listed, but nevertheless have played a crucial role in the development and execution of the projects.

Strung (Melbourne, 2013)

Producer: Arts Centre Melbourne (Riverside Live Program)
Concept and design: Tanja Beer
Performer: Sabrina D'Angelo
Active Scenographers: Jennifer Tran and Ryan Foote
Photos: Gisela Beer
Film: Sam Hoffmann

Strung (Cardiff, 2013)

Producer: World Stage Design
Concept and Design: Tanja Beer
Performer: Elyssia Sasaki
Active Scenographers: Garrett Levine and Donyale Werle

Strung (London, 2013)

Producer: Collisions (Royal Central School of Speech and Drama)
Concept and Design: Tanja Beer
Performer: Christina Kapadocha
Active Scenographers: Ella Marie Fowler, Jacquie Holland and Natalie Jackson
Sound: Harry Webber
Photos and film: Alex Blake
Thank you to: Lisa Woynarski, Peter Bingham, Susanne Page and David Lam

This Is Not Rubbish Craft Circle (London, 2013/2014)

Producer: Green Week (Royal Central School of Speech and Drama)

Concept and Design: Tanja Beer

Co-facilitator: Susanne Page

Craft Circle participants: Susanne Page, Zoe White, Jolanta Klatt, Heather Ashley, Ella Marie Fowler, Jacquie Holland and Lucy Latham

Models: Elisabeth Murphy, Christina Kapadocha and Jimmy

Photoshoot assistance: Heather Ashley, Susanne Page, Jolanta Klatt, TSD and Lynn Ahrens

Photos: Alex Blake

The Living Stage (Castlemaine, 2013)

Producers: Castlemaine State Festival

Performances: *Produce* (Born in a Taxi & CreateAbility), *The Long Table Feast* and *Garden Chef*

Design: Tanja Beer with community

Assistant Artist: Melinda Rodnight

Production Assistant: Milton Perks

Growing: Hamish MacCallum, Sas Allardice and the Castlemaine Community

Documentary filmmakers: Sam Hoffman and Jim Coad

Publicity: Sassy Red Publicity

Photos: Gisela Beer

Thank you to all those who've kindly given their time, ideas and resources to the project: Geoff Heard, Jika MacCallum, Allie Hanly, Sam Downing, Grace McCaughey, Rohan Souter, Lucy Souter, John Willis, Judy & Philip Hopley, Margaret Griffin, Lena Mitchell, Dorothy Wright, Elizabeth Telford, Daniel Williams, Jack Telford, Henri Hamlin, Ruben Evens, Francis Hamlin, Vanessa Case, Sequani MacCallum Telford, Arwen MacCallum, Arion MacCallum and Oliver Harris

Thank you to Pozible supporters and grant-funders: Regional Arts Victoria, Castlemaine State Festival and Growing Abundance.

The Trans-Plantable Living Room (Cardiff, 2013)

Producer: World Stage Design

Performance: Plantable Research Collective (Lisa Woynarski, Bronwyn Preece and Megan Moe Beitiks)

Lead Designer: Tanja Beer

Assistant Designer: Erini Gregoriades

Growing and design: Sam Holt and the Riverside Community Allotments (A project that provides a place where people can learn how to grow food in a sociable and supportive atmosphere)

Documentary film-maker Film: Rabab Ghazoul

Producers and promoters: Sarah Jayne Leigh and Mawgaine Tarrant-Cornish

Photos: Nigel Pugh, Valeria Pacchiani and Flor Dias

Acknowledgements: Becca Clarke of Green City Events, Michele Fitzsimmons of Edible Landscaping, Polly Reichelt and Adamsdown Community Garden, Alistair Rowan and Field Days Organics (Innovate Trust) and Vegalive Aquaponics. Katie Jones of Federation of City Farms and Community Gardens, People's Collection Wales

Interviewees: Anne Bateman, Anon, Richard Berry, Gordon David Clarke, Sophie Durnan, Edith England, Michele Fitzsimmons, Stephen Garrett, Sam Holt, Jason Horn, Kate Knowles, Yvan Maurel, Lewis Mottrum, Monica Nobriga, Helen Reardon, Mariusz Rogacz, Tim Stevenson, Stephen Watts, Dafyd Williams

Thanks to all those who kindly gave their time, ideas and resources to the project: Claire Bracegirdle, Frances Smith-Williams, Eva Krasenska, Owen Evans, Ian Garrett and CSPA, Gisela Beer, Manuela Franken, Freya Stanley-Price, Rosie Strickland, Kit Jones, Allan Shepherd, Ariana Jordao, Harry Giles, Emma Williams and everyone at World Stage Design

Special thanks to Wefund supporters and grant-funders: Artist's Project Earth, World Stage Design, Edes Foundation and Our Food

Uprooted (Glasgow, 2015)

Producer: Eco Drama

Director/Devisor: Emily Reid

Lead Designer: Tanja Beer

Growing: Katie Lambert (with Corpus Christi Primary, St Vincent's Primary, Balornock Primary and Aultmore Park Primary)

Assistant Designer: Mona Kastell

Set Building: Colin Myles and Alistair Jack

Performers/Devisors: Caroline Mathison, Greg Sinclair and Marta Masiero

Dramaturge: Katrina Caldwell

Special Thanks to our funders: Esmée Fairbairn Foundation, The Robertson Trust Creative Scotland, Ernest Cook Trust, Glasgow Arts and Glasgow City Council

The Bower Stage (Armidale, 2016)

Producers: EcoArts Australis Inc & Black Gully Music Festival with support from New England Regional Art Museum (NERAM) and Australian Government Festivals Australia. The Bower Stage was part of *Ephemera*, curated by EcoArts Australis Inc.

Lead designers: Tanja Beer, Ashlee Hughes and Simon Mellor

Eco-drama performance team: Julie Collins, Camille Dunsford and Katy Walsh with students from Armidale High School and Duval High School.

Key contributors: Armidale Tree Group and Armidale Community Garden

Ephemera artists: Andrew Parker, Gabi Briggs, Amy Hammond, Greer Taylor and Laszlo Szabo (Social Ventures Media)

Producing team: Cherene Spendelove, Andrew Parker and Dave Carr

Thank you to: The Backtrack Boys, Armidale Steiner School and NERAM.

Photos: Laszlo Szabo

Running Wild (Frankston North, 2016)

Producer: Polyglot Theatre

Project director/lead teaching artist: Kate Kantor

Designer/researcher/teaching artist: Tanja Beer

Park ranger/teaching artist: Richard Faulkner

Indigenous elder/teaching artist: Linda Mullett (Kurnai)

Percussionist/teaching artist: David Joseph

Filmmaker: Jason Heller

Design assistant/teaching artist: Jennifer Tran

Producing support: Lis Blake

Production manager: Lexie Wood

Teaching staff: Michael Shanks, Michelle Cottick and Nathan Wilson

Thank you to: Students and staff of Mahogany Rise Primary School and Monterey Secondary College

Photos: Sarah Walker

The Living Stage NYC (New York, 2017)

Producers: Superhero Clubhouse and University Settlement

Lead designers: Tanja Beer and XDEA Architects

Team: Alison Fleminger, Azizi Curtis, Carolyn Chadwick, Cassiope Sydoriak, Chris Ignacio, Dawn Crandell, Gabi Nail, Jeremy Pickard, Karen Ng, Katherine Pena, Lanxing Fu, Maria Delgado Jones, Mark Pundzak, Milton Perks, Nic Adams, Olivia Johnson, Pia Guilliatt, Scott Johnson, Sergey Pigach, Tanja Beer, Tonya Ingerson and Trevor New

Photos: Dylan Lopez and Milton Perks

Thank you to: Meltzer Seniors and the Lower East Side NYC community

The Living Stage (Lorne, 2018)

Producers: Lorne Sculpture Biennale with support from the Thrive Research Hub (Melbourne School of Design, The University of Melbourne), the Great Ocean Road Coast Committee and Graham Blashki & Evelyn Firstenberg

Lead designers: Tanja Beer and Ashlee Hughes

Community support & growing team: Helen Smith, Anne Nadenbousch, Colin Leitch, Sue Grant and Grace Nicholls with the Lorne Kindergarten

Assistant Designers: Pia Guilliatt and Milton Perks

Set builder: Tim Denshire-key

Performers: Mountain Grey, Randall Forsyth, Helen Duncan & Sofie Burgoyne

Plants donated by: Batesford Nursery, Bushland Flora, Flinders Nursery, Tavistock Nursery, Tree Growers Advanced (TGA), Warners Nurseries and Rhodo Glen Nurseries

The Living Pavilion (Melbourne, 2019)

The Living Pavilion was a co-production and collaboration with Clean Air and Urban Landscapes Hub (CAUL) of the National Environmental Science Program, THRIVE Hub (Faculty of Architecture, Building and Planning), the New Student Precinct of the University of Melbourne's Parkville campus, and CLIMARTE's ART+CLIMATE=CHANGE 2019 Festival. *The Living Pavilion*'s major horticultural and design partners were Australian Institute of Landscape Architects (AILA) and Ecodynamics. Other partners of *The Living Pavilion* included Next Wave, Place Agency, BILI Nursery, 226 Strategic, Graduate Student Association (GSA), The Living Stage, Garawana Creative, Melbourne Sustainable Society Institute and Sustainability Team @Unimelb.

Producers: Cathy Oke, Tanja Beer & Zena Cumpston

Associate Producer: Jeremy Taylor (New Student Precinct)

Lead researcher: Zena Cumpston

Indigenous advisory team: Charles Solomon, Dean Stewart, Zena Cumpston, Mandy Nicholson, Maddison Miller, with additional support from Greenshoots Consulting, Murrup Barak and Wilin Centre, Jefa Greenaway, CAUL Hub's Indigenous Advisory Group and Rueben Berg

Original concept (The Living Stage) and lead designer: Tanja Beer

Assistant designers: Pia Guilliatt, Camille Greenfield, Zongjing Yu & Zachariah Dahdoule

Design coordinator: Ashlee Hughes

Contributing designers: Zena Cumpston, Steph Beaupark

Signage: Illustrations by Dixon Patten of Bayila Creative, research and words by Zena Cumpston, design and production by 226 Strategic

Graphic design (program): Dixon Patten of Bayila Creative (Principal) and Rachel Pirnie (UoM)

Program partnerships and events manager: Cathy Oke

Programming and event assistant managers: Jeremy Taylor, Rachel Iampolski, Skylar Lin, Amelia Leavesley, Marley Holloway-Clarke, Anita Spooner and Paris Paliouras

Communications managers: Isabel Kimpton, Nicole Mustedanagic, Leah Hyland, Sophie Hill, Cathy Oke, Kiah McCarthy and Alice Tovey

Communications strategy and social media: Isabel Kimpton

Horticulture design and propagation team: Nick Somes, Jeff Beavis, Randall Wee, Adrian Gray, Charles Solomon, Zena Cumpston, Chelsie Davies, Kate Hogan

The Living Pavilion student ambassadors: Gabrielle Margit Lewis, Jane Chen, Chelsea Matthews, Victoria Tabea Seeck, Lucia Marie Amies, Mimmalisa Trifilò and Rachel Lampolski

Horticulture coordinator: Jenny Pearce

Site and plant maintenance: Milton Perks

Wayfinding design and materials: Helaine Stanley and Andrew Hubbard, 226 Strategic

Research team: Cristina Hernadez Santin, Tanja Beer, Zena Cumpston, Rimi Khan, Luis Mata, Kirsten Parris, Christina Renowden, Rachel Iampolski, Leila Farahani, Eugenia Zoubtchenko and Blythe Vogel

Financial management: Angela Bruckner and Siouxzy Morrison

Site support: Tevita Lesuma, Suzanne Griffin, Louise Ryan (Graduate Student Association), Rob Oke and Lewis Mcleod

Information booth: Kay Oke, ART+CLIMATE=CHANGE 2019 festival crew

Soundscape: Mark Pollard, Alex Beck, Trev Dunham, Lachlan Wooden and The Faculty of Fine Arts and Music (Interactive Composition students)

Co-design creative development workshop contributors: Judith Alcorn, Margaret Bakes, Steph Beaupark, Tanja Beer, Chelsie Davies, Christina Chiam, Zena Cumpston, Harriet Deans, John Delpratt, Marita Dyson, Michael Ford, Lisa Godhino, Adrian Gray, Cris Hernandez, Joe Hurley, Leah Hyland, Sophie Jackson, Ryan Jefferies, Bronwyn Johnson, Kate Kantor, Alex Kennedy, Amelia Leavesley, Meredith Martin, Luis Mata, Patrick Mercer, Sue Murphy, Cathy Oke, Jenny Pearce, Eleanor Percival, Anne-Marie Pisani, Ian Shears, Robert Snelling, Jeremy Taylor, Alice Tovcy and Katie West

University of Melbourne and New Student Precinct contributors: Georgie Meagher, Alex Kennedy, Mal Abley, Chris Frangos, Tim Uebergang, Dominic Napoleone, Dani Norman, Danny Butt and Mark Gillingham

The Living Pavilion project partners and collaborators would like to acknowledge the Traditional Custodians of the land and waterways on which the project took place, the Wurundjeri peoples of the Woi Wurrung language group, part of the greater Eastern Kulin Nations. We pay our respects to Wurundjeri Elders, past, present and emerging. We honour the deep spiritual, cultural and customary connections of the Traditional Custodians to the landscape and ecology of the land on which *The Living Pavilion* is located. We acknowledge that this land, of which we are beneficiaries, was never ceded and endeavour to reflect and take consistent action to address this harmful circumstance. We are especially grateful for the contributions of many First Peoples involved in our project and their generosity to share their culture and knowledge with us. Without them, *The Living Pavilion* would simply not have been possible.

Index[1]

A

B

[1] Note: Page numbers followed by 'n' refer to notes.

T. Beer, *Ecoscenography*, https://doi.org/10.1007/978-981-16-7178-4

The manufacturer's authorised representative in the EU is Springer Nature Customer Service Centre GmbH, Europaplatz 3, 69115 Heidelberg, Germany. If you have any concerns regarding our products, please contact ProductSafety@springernature.com

Printed and bound by CPI Group (UK) Ltd, Croydon, CR0 4YY

30/09/2026

02225321-0005